Living with the Lake Erie shore

Living with the shore

Living with the Lake Erie shore

Charles H. Carter   William J. Neal   William S. Haras   Orrin H. Pilkey, Jr.

Sponsored by the National Audubon Society™

Duke University Press   Durham and London

The National Audubon Society and Its Mission

In the late 1800s, forward-thinking people became concerned over the slaughter of plumed birds for the millinery trade. They gathered together in groups to protest, calling themselves Audubon societies after the famous painter and naturalist John James Audubon. In 1905, thirty-five state Audubon groups incorporated as the National Association of Audubon Societies for the Protection of Wild Birds and Animals, since shortened to National Audubon Society. Now, with more than half a million members, five hundred chapters, ten regional offices, a twenty-five million dollar budget, and a staff of two hundred seventy-three, the Audubon Society is a powerful force for conservation, research, education, and action.

The Society's headquarters are in New York City; the legislative branch works out of an office on Capitol Hill in Washington, D.C. Ecology camps, environmental education centers, research stations, and eighty sanctuaries are strategically located around the country. The Society publishes a prize-winning magazine, *Audubon*, an ornithological journal, *American Birds*, a newspaper of environmental issues and Society activities, *Audubon Action*, and a newsletter as part of the youth education program, *Audubon Adventures*.

The Society's mission is expressed by the Audubon Cause: to conserve plants and animals and their habitats, to further the wise use of land and water, to promote rational energy strategies, to protect life from pollution, and to seek solutions to global environmental problems.

National Audubon Society 950 Third Avenue New York, New York 10022

Research for and publication of this volume was subsidized in part by the George Gund Foundation.

Publication of the various volumes in the Living with the Shore series has been greatly assisted by the following individuals and organizations: the American Conservation Association, an anonymous Texas foundation, the Charleston Natural History Society, the Coastal Zone Management Agency (NOAA), the Geraldine R. Dodge Foundation, the William H. Donner Foundation, Inc., the Federal Emergency Management Agency, the George Gund Foundation, the Mobil Oil Corporation, Elizabeth O'Connor, the Sapelo Island Research Foundation, the Sea Grant programs in New Jersey, North Carolina, Florida, Mississippi/Alabama, and New York. The Fund for New Jersey, M. Harvey Weil, and Patrick H. Welder, Jr. The Living with the Shore series is a product of the Duke University Program for the Study of Developed Shorelines, which is funded by the Donner Foundation.

Library of Congress Cataloging-in-Publication Data
Living with the Lake Erie shore.
(Living with the shore)
"Sponsored by the National Audubon Society."
Bibliography: p.
Includes index.
1. Shore protection—Erie, Lake, Region. 2. Coastal zone management—Erie, Lake, Region. 3. Coasts—Erie, Lake, Region. I. Carter, Charles H. (Charles Henry) II. Series.
TC225.E6L58 1987   333.91'7'097712   87-5398
ISBN 0-8223-0678-6 ISBN 0-8223-0741-3 (pbk.)

Contents

Figures and tables

Figures

Tables

Preface

There is something very special about the Lake Erie shore, where the air, the water, and the land meet. A stroll along a beach or a quiet moment on a rocky promontory as the sun sets renews one's spirit in a way difficult to describe. However, haphazard development of the Lake Erie coastal zone threatens the very reason we are attracted to the coast: to escape, if for but a short while, the cares and worries of our modern, fast-paced society. Along with this haphazard coastal development come ongoing accounts of threatened or destroyed property and the human anguish that accompanies the damage and destruction. Moreover, much of the loss can be traced to structures that initially were built to protect the shore from flooding and erosion. These structures are causing the following effects: beaches are becoming narrower and less continuous; and erosion rates are becoming more variable, leading to a more irregular shoreline. Also, the shore is beginning to take on an armored appearance as if the inhabitants were getting ready to repel an invasion.

We do not claim to know all the answers, but as geologists we know that erosion and flooding are likely to continue and will worsen if the lake level rises. We strongly feel that we need a longer-range, more ordered approach to coastal development along the Lake Erie shore. Coastal problems are getting worse, and if we are going to conserve the shore for future generations, we need to begin to work on these problems now.

This book is directed to the coastal dweller, coastal developer, land-use planner, and those concerned with the future of Lake Erie. Our goal is to retell a little of the history, discuss coastal processes, define the kinds of hazards faced by coastal residents and property owners, and indicate how to reduce the impacts of those hazards. Knowledge about normal lake processes can lead to more informed choices.

The present volume is part of the "Living with the Shore" series being published by Duke University Press, which will eventually cover most coastal states. As an umbrella book to the series, Duke Press has reprinted *The Beaches Are Moving: The Drowning of America's Shoreline* (1983) by Wallace Kaufman and Orrin H. Pilkey, Jr., including an updated appendix. Each book is written by local experts. Dr. Charles Carter, University of Akron, Akron, Ohio, was invited to guide this volume because of his many years of work on the Lake Erie shore as a coastal geologist, including ten years with the Ohio Geological Survey. Dr. Carter has authored numerous scientific publications on coastal processes. Mr. William Haras of the Canada Centre for Inland Waters, Burlington, Ontario, was invited to provide a summary of the available information for the Canadian shore. Mr. Haras, an engineer/hydrographer/planner, has thirty years of experience on the Great Lakes. He has served on the International Joint Commission (IJC) Great Lakes Water Level Studies and has authored numerous publications on the Canadian shore. To our delight, these two scientists agreed to accept the time-demanding task of summarizing a great deal of information into a useful layman's guide. We participated both as editors and contributing authors. Dr. William J. Neal is a geologist at Grand Valley State College, Allendale, Michigan, working on Great Lakes topics through the auspices of that school's Water Resources Institute. Dr. Orrin Pilkey, Jr., James B. Duke Professor of Geology, Duke University, Durham, North Carolina, is a coastal geologist of long experience, particularly with barrier coasts. He heads the Program for the Study of Developed Shorelines, and is a coauthor with Dr. Neal of *Coastal Design: A Guide for Builders, Planners, and Homeowners* (1983), which outlines coastal construction principles that can mitigate some coastal hazards.

The successful completion of even a small book is no simple task. This volume came about, in part, because of the interest and financial support of the George Gund Foundation of Cleveland, Ohio. We extend a sincere thanks to the Gund Foundation for their support and interest in public environmental education. The overall coastal book project is an outgrowth of initial support from the National Oceanic and Atmospheric Administration (NOAA) through the Office of Coastal Zone Management. The initial project was administered through the North Carolina Sea Grant Program. More recently it has been generously supported by the Federal Emergency Management Agency (FEMA). Without FEMA support the series would not have proceeded this far. However, the conclusions presented herein are those of the authors and do not necessarily represent those of the supporting agencies. We have been helped by many individuals and organizations in the preparation of this book. In particular we thank the following people for reviewing parts of the manuscript: Martin Jannereth (Michigan Department of Natural Resources), Richard Bartz (Ohio Department of Natural Resources), Shamus Malone (Pennsylvania Department of Environmental Resources), Aram Terchunian (New York Department of State), Denton Clark (U.S. Army Corps of Engineers), and David Strelchuk (Ontario Ministry of Natural Resources). Dr. Robert P. Bukata was especially helpful in discussion reviews with Bill Haras concerning the north shore of the lake. John Shaw and Dr. Reid Kreutzwiser also provided valuable input. G. M. Kutz and Laura Boone helped prepare the strip maps and associated data from the Canadian side. James Kerwin (*Detroit News*), Henry Harvey (*Toledo Blade*), and William Ironside (*London Free Press*) were especially helpful in finding photographs. Particularly helpful organizations were the U.S. Army Corps of Engineers, Detroit and Buffalo district offices; Michigan Department of Natural Resources; Ohio Department of Natural Resources; Pennsylvania Department of Environmental Resources; New York Department of State; NOAA Great Lakes Environmental Research Laboratory, Environment Canada, Ontario Ministry of Natural Resources, and the International Joint Commission.

We owe a debt of gratitude to many individuals for support, ideas, encouragement, and information. Doris Schroeder has helped us in many ways as Jill-of-all-trades over a span of more than a decade and more than a dozen books. Duke University

Press staff compiled the index for this volume. The original idea for a coastal book (*How to Live with an Island* [1972]) was that of Pete Chenery, then director of the North Carolina Science and Technology Center. Richard Foster of the Federal Coastal Zone Management Agency supported the project at a critical juncture. Because of his lifelong commitment to land conservation, Richard Pough of the Natural Area Council has been a mainstay in our fund-raising efforts. Myrna Jackson, of the Duke Development Office, and the President's Associates of Duke University have been most helpful in our search for support.

Mike Robinson, Jane Bullock, and Doug Lash of the Federal Emergency Management Agency have helped us chart a course through the shifting channels of the federal government. Richard Krimm, Peter Gibson, Dennis Carroll, Jim Collins, Jet Battley, Melita Rodeck, Chris Makris, and many others have opened doors, provided maps, charts, and publications, and generally helped us through the Washington maze.

Last, but not least, we extend our thanks to Bette Weerstra, Traci Baker, Melissa Greer, and Kelly Thomas for typing and retyping the manuscript; Barbara Gruver, Tonya Clayton, Leslie Droege, Lynne Claflin, and Don Powell for drawing the line illustrations and maps; and Nancy Margolis for editorial contributions.

As in any effort of this sort, we were helped by many people who live and work along the shore—again, so many we cannot list them all. We are grateful for their cooperation and for the insight and concern we have gained from them. We dedicate this work to them and to all who enjoy Lake Erie's shores, today and tomorrow.

William J. Neal
Orrin H. Pilkey, Jr.
Series Editors

1   A coastal perspective

Fresh water, sunsets, beaches, waterfowl, port towns and cities, walleye, lake ferries and freighters, the islands, surf, Cedar Point, sailboats, Point Pelee, lake homes, Presque Isle, vineyards and wineries: these words are commonly used in describing the Lake Erie shore, truly a major natural resource (figs. 1.1A and 1.1B). The people of two great countries enjoy these amenities, but all is not well along this Great Lake. Coastal hazards, particularly shoreline erosion and flooding, have set off an environmental, economic, and political chain reaction that is yet to be contained.

The attractiveness of the lakeshore combined with the affluence of our society has led to increased development of the shore land, which in turn leads to greater storm damage and economic loss every time the lake level is high (figs. 1.2 and 1.3). Two great forces, human development and natural wave-topped lake levels, are making the shoreline a battleground. Property owners see their dwellings and dreams as the potential victims of the battle.

A U.S. Army Corps of Engineers study, for example, shows that between 1972 and 1976 about $58 million were spent on shore erosion protection, and about $90 million in property losses were incurred along the U.S. shore (reference 54, appendix B). On the northern, Canadian shore, costs for one year from November 1972 to November 1973 included: shore damage, CN$2.9 million; remedial works, CN$1.9 million; and lost land value, CN$2.6 million (reference 50, appendix B). The greater U.S. losses reflect the greater development of the lakeshore along the U.S. side, particularly along the western half of the lake where the elevation of the shore land is low, and where shore materials are more easily eroded. The data also reflect a period (early 1970s) of record-setting lake levels that, on the basis of historic trends, were expected to fall, but that persisted into the mid-1980s to again set record levels!

The point is that lake level and storms, the natural dynamic processes that cause erosion and flooding, are forces that lakeshore property owners must reckon with and plan for. Serious shoreline problems can persist for several years, and will recur at frequent, but irregular intervals.

The storm of November 13 and 14, 1972, is a case in point. The severe damage resulted from the combination of unrestricted development, storm intensity, and high lake level. A north-northeast wind, which reached a speed of 60 knots (69 mph), blew for two days directly down the long axis of the lake. The wind generated waves at least 12 feet high and caused a storm surge (temporarily high lake level) more than six feet above the lake's average November level. This water level was about four feet above the previous record-high lake level set in that same month of November 1972! Erosion and flood damage caused by the waves and high water led to the declaration by the U.S. president of a major disaster area for most of the Ohio and Michigan shore.

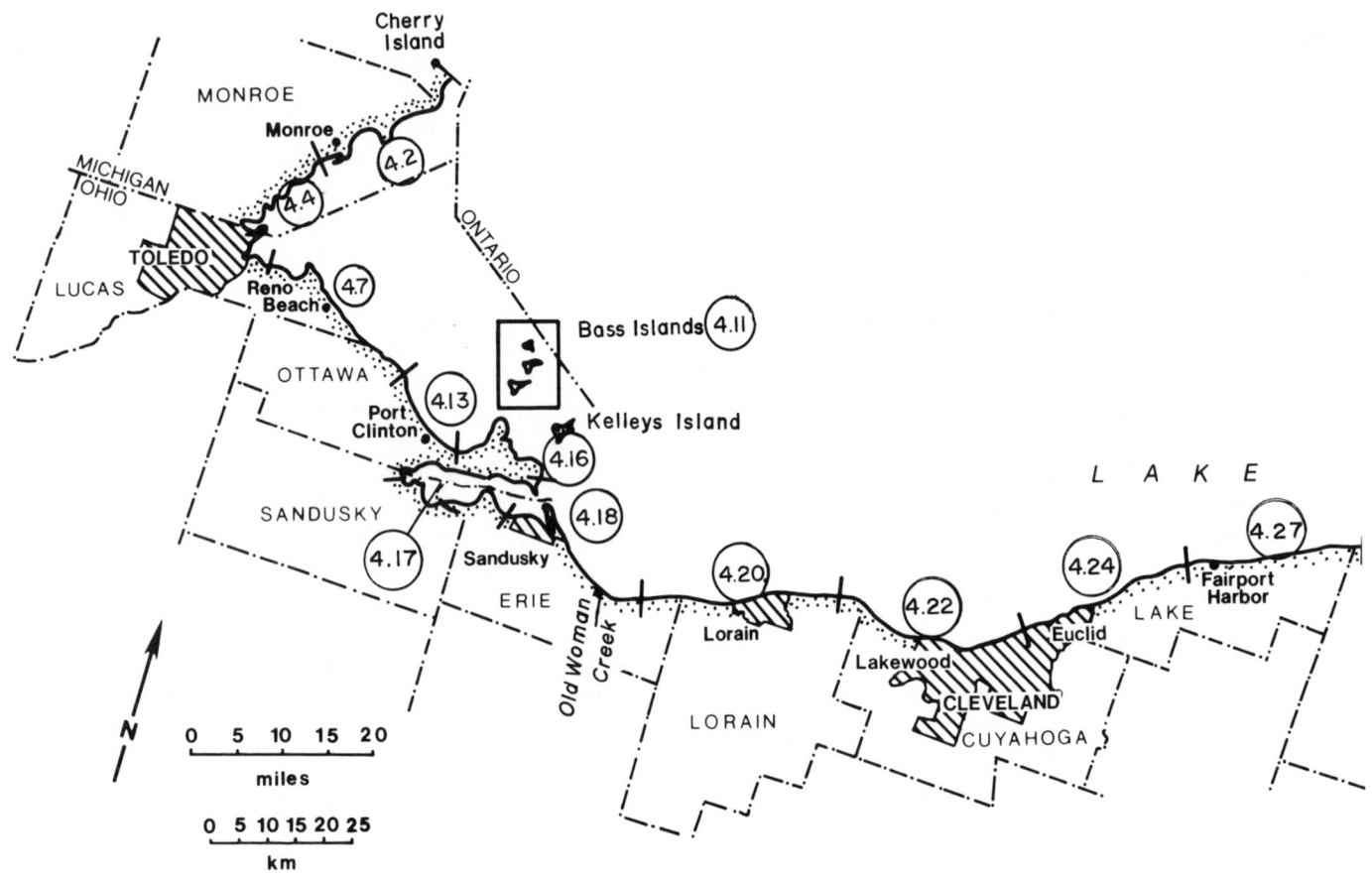

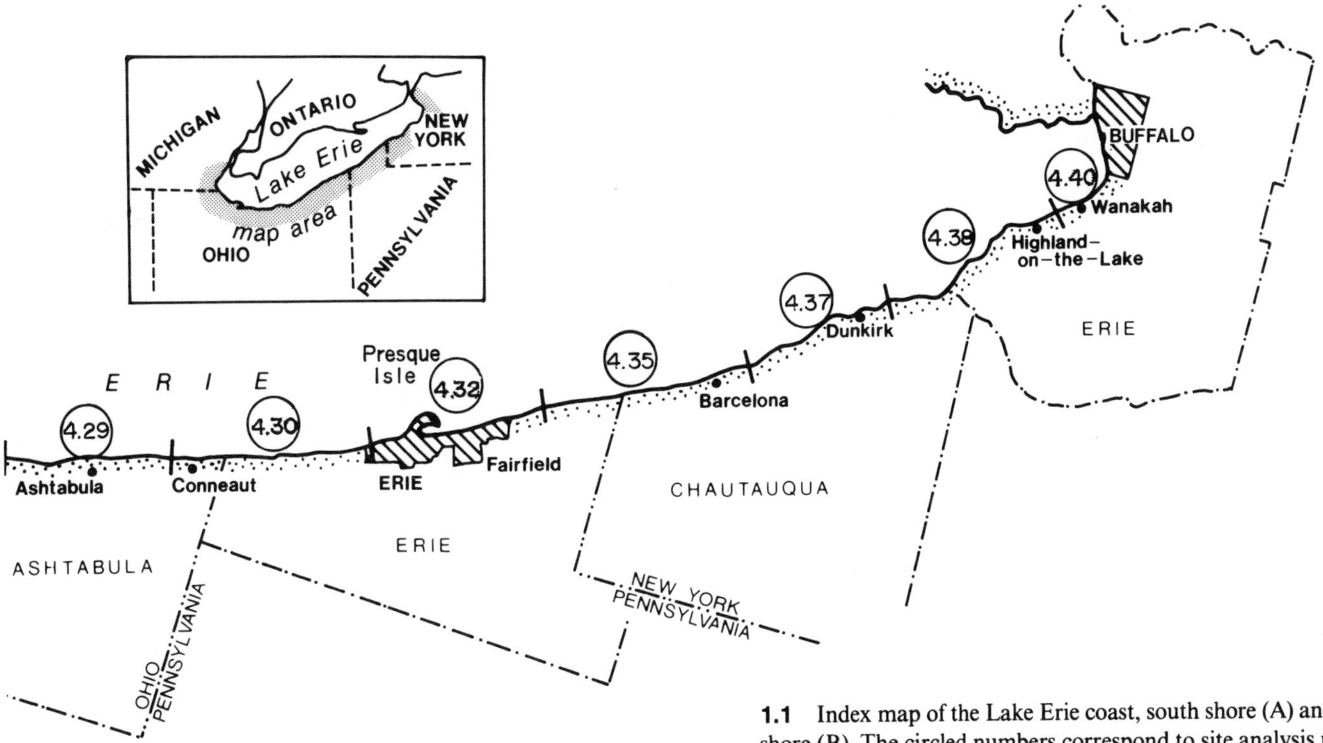

MICHIGAN
ONTARIO
NEW YORK
Lake Erie
map area
OHIO
PENNSYLVANIA

BUFFALO

4.40
Wanakah

4.38
Highland–
on–the–Lake

ERIE

4.37
Dunkirk

4.35

Barcelona

Presque
Isle

E R I E

4.32

4.29

4.30

ERIE

Fairfield

CHAUTAUQUA

Ashtabula

Conneaut

ERIE

ERIE

ASHTABULA

OHIO
PENNSYLVANIA

NEW YORK
PENNSYLVANIA

1.1 Index map of the Lake Erie coast, south shore (A) and north shore (B). The circled numbers correspond to site analysis maps (chapter 4).

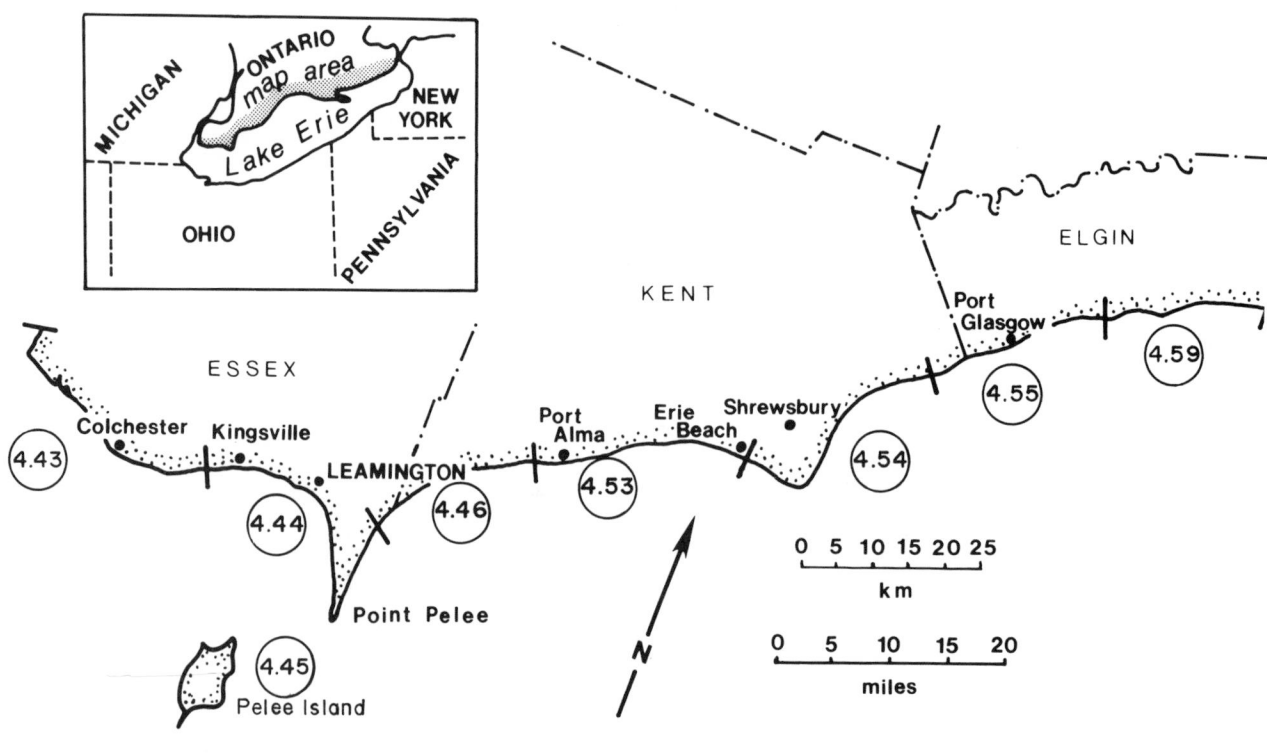

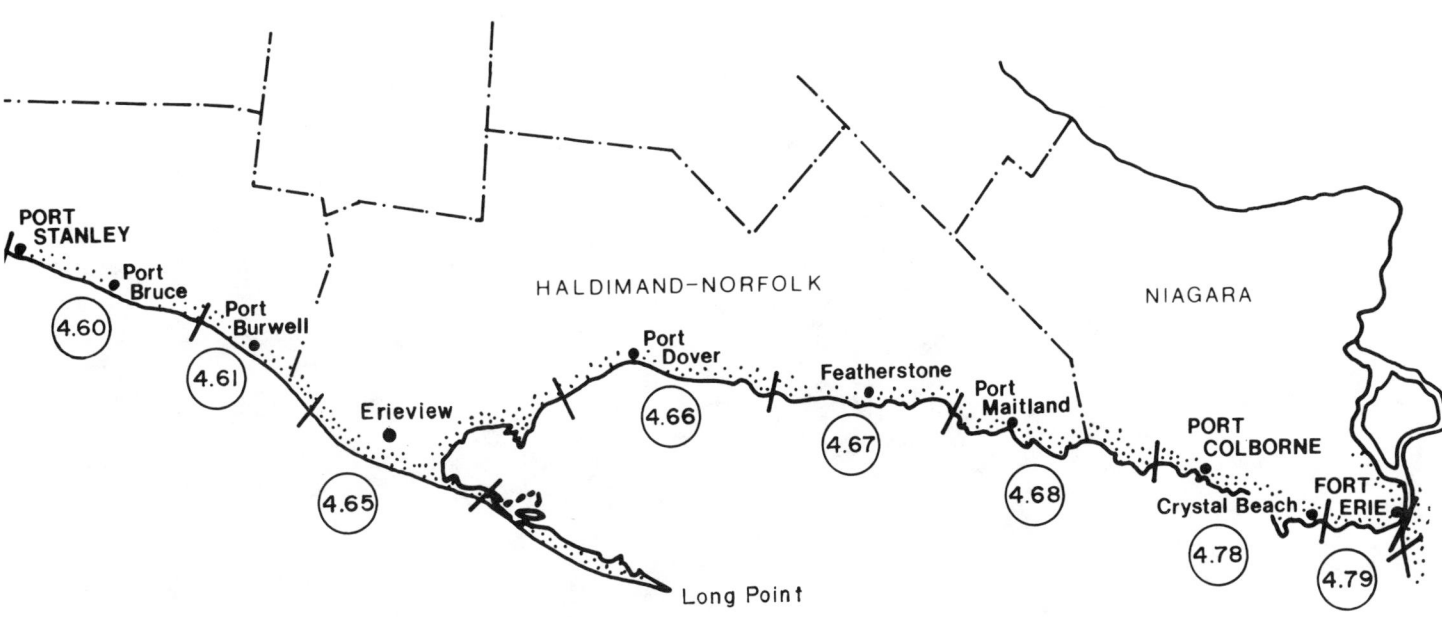

PORT
STANLEY

Port
Bruce

Port
Burwell

4.60

4.61

Erieview

HALDIMAND-NORFOLK

Port
Dover

Featherstone

Port
Maitland

NIAGARA

4.65

4.66

4.67

PORT
COLBORNE

4.68

Crystal Beach

FORT
ERIE

4.78

4.79

Long Point

E R I E

In Ontario the same November storm sent high waves crashing into Canadian shorelines already menaced by rising water. Wide areas of Southwestern Ontario were flooded. Thousands of people were removed from low-lying shore areas.

On Pelee Island, eight miles from the mainland, cottages and homes were swept off their foundations as a tormented Lake Erie smashed dikes and spilled floodwaters over nearly half the island of 300 year-round residents.

"There's no use pumping it out," sighed Pelee Township reeve William Bartja as he surveyed the destruction to island farmlands, reputed to be among the best in the province. "It's all level with Lake Erie now. . . ."

The storm died down the next day but it vividly brought to public attention the increasing seriousness of the lake's rise. (The *London Free Press*, March 14, 1973.)

The senior author remembers the storm well, since he was living in a rented house at the west end of the lake while studying the effects of storms (reference 44, appendix B). Many homes built close to the lake in low-lying areas were totally destroyed by wave impact, leaving nothing but piles of rubble. Even some sturdier homes and homes built on higher ground sustained major structural damage from the waves, storm surge, and sand

1.2 The fireplace is all that remains of a former Long Point, Ontario, cottage destroyed by Lake Erie's high waters. Photograph by Bill Ironside, courtesy of the *London Free Press*.

1.3 November 1972 storm damage at Reno Beach, Ohio. Photograph courtesy of the *Toledo Blade*.

transported by the water (figs. 1.3 and 1.4). Along the Cedar Point chaussée, the beaches, dunes, and trees that had been considered relatively good natural protection turned out to be insufficient against the force of the storm (fig. 1.5). The beaches and dunes were severely eroded. In places, the road was under-

1.4 Damaged houses at Sand Beach, Ohio, after the November 1972 storm.

1.5 Post-storm cleanup along the Cedar Point, Ohio, chaussée, November 1972.

cut. Trees were downed or undercut, and tremendous quantities of sand were washed over and through the dunes by the storm waves.

These problems were just a portent of things to come. In the spring of 1973 record-high lake levels and spring storms battered the lakeshore (fig. 1.6). For many coastal residents it was a true crisis. Construction engineers and construction material businesses worked practically nonstop to try to keep up with the demand of the lakeshore property owners for construction of new or rebuilt shore protection structures. Federal and state governments processed hundreds of claims for low-interest disaster loans and applications for shore protection structure permits. Signs saying "For Sale" or "Fill Wanted" were posted as owners either tried to move or attempted to protect their property.

Afterward the lake level fell below the records of 1973, but it stayed at least a foot above the long-term average for the next 12 years. During this time there was an uneasy truce between the

1.6 April 1973 storm strikes Detroit Beach, four miles north of Monroe, Michigan. The house lakeward of the seawall was eventually battered to pieces by the waves. Houses in back of the wall were also damaged by flooding and waves. Photograph courtesy of the *Detroit Free Press*.

lakeshore property owners and the lake. This truce ended in 1985 when the lake again reached record levels and fall storms again took their toll on lakefront property.

On December 2, 1985, strong southwest winds blowing across a December record-high lake level generated 12-foot-high waves and a record storm surge elevation of 580.6 feet. This record high came only seven years after the previous record high, which occurred during an April 1979 storm. Unlike the effects of the 1972 storm, the damage this time was concentrated at the east end of the lake. The small communities of Hoover Beach, Woodlawn, and South Buffalo in the Buffalo area were hardest hit (fig. 1.7).

The December 1985 storm caused $3 million in losses due to erosion and flooding. At least 12 houses were completely destroyed as most shore protection structures could not hold back the onslaught of waves and storm surge. Lakeshore buildings were smashed and flooded. Protective structures were damaged or destroyed. Federal and state officials, quickly on the scene to assess the damage, could give no promise of assistance to the unfortunate landowners. No federal disaster was declared by the president, so no low-interest loans were available to rebuild and repair damaged structures. Owners without insurance were faced with both personal loss and dollar loss. The same sad and familiar questions were being raised. Where can we go? How am I going to cover my losses? Could I have avoided this disaster?

In spite of such history, lakeshore owners continue to build

1.7   December 1985 damage and flooding in the Hoover Beach area at Hamburg, New York. Photograph by Ronald J. Colleran, courtesy of the *Buffalo News.*

shore protection structures that they hope will protect their property from erosion and flooding. Landowners, many of whom bought their homes during the low-lake-level years of the 1960s, keep a close eye on lake level because they now realize that a relatively small increase in water level, particularly when the lake is already high, can have enormous consequences. Along much of the Erie shore no beach remains, and the proliferation of seawalls, groins, and breakwaters, as well as remnants of various structures along the more densely populated stretches of the shore give the appearance of an army obstacle course. The price of protecting private homes and property along the shore has been the loss of beaches, a public resource.

Vernon and Mary Triplett of Vermilion, Ohio, provide a good example of lakeshore owners who have withstood the onslaught of Lake Erie storms since 1948. In 1948 when the Tripletts moved into their home, they were faced with a largely unprotected, 32-foot-high clay bluff. A seawall that had been built in about 1944 had failed. Using the old seawall as a footer, they reconstructed the wall by building forms and mixing cement from the sand and gravel on the beach during evenings, weekends, and holidays. One summer they acquired railroad ties and with hired help constructed several riprap-filled cribs along the eastern part of their property. In addition, stone blocks from a nearby quarry were emplaced. The lake level had been rising since 1948 and reached historic record levels in 1952. The Tripletts' shore protection worked reasonably well and erosion from wave splash and slope failure was confined mainly to the unprotected upper bluff. It was not until the late 1950s that the lake level dropped to near its long-term average, which gave the Tripletts a brief respite from the maintenance of their shore protection structures. In this first ten-year period they had spent about $10,000 for shore protection materials and construction, not even counting their own hard labor!

In the late 1950s the Tripletts designed 850-pound concrete prisms that were linked together with steel cable through rings. These structures worked well when the lake was low and as long as there were no major storms. From 1958 to 1968 the Tripletts did little work on the structures as lake level was for the most part below the long-term average. In 1968 they had fill placed in front of their seawall to form a beach, but following the record-high lake levels and major lake storms of 1972 and 1973 even these structures were torn loose and lost. Nonetheless they were able to obtain a Small Business Administration loan for $11,000 that largely covered their expenditures of about $13,000 for culvert pipe. The pipe was placed on end and filled with concrete to replace, in part, their badly damaged seawall. In 1985, with the lake again at record levels (it never dropped much after 1972, a phenomenon that has confounded some of the lake level experts), they were again confronted with major repair problems as their culvert pipe sections broke and began toppling into the lake.

The Tripletts have been relatively successful at protecting their lakeshore property. Nevertheless, their bluff has retreated about 25 feet since 1948. This is less than one foot per year, but the bluff is now within about 20 feet of their home in spite of all the protective works, the untold tons of fill that were handled, dumped, pushed over, graded, and planted, and the trees that were planted and encouraged to grow in the hope that the roots would help to hold the land in place. In total, they have spent more than twice the original cost of their property for shore protection, not to mention their hard physical labor and psychological distress. If they had to do it over again, they would; but they would start with a house at least 100 feet away from the lake (not 44 feet) from which to watch the water and the sunsets. Unlike people, lakeshore erosion is untiring.

The potential for property damage and loss, and the threat to the personal well-being of coastal dwellers and property owners is what this book is about. Our goal is to identify the risks, so that coastal dwellers can avoid or reduce them through prudent choices. To do this we consider how we've come to be on the shore and the lake's origin (chapter 1), processes within the lake (chapter 2), and the impact of human activity on the shore (chapter 3). Once these fundamentals are in hand, we evaluate the risk factors, focusing on coastal classification so property owners can evaluate their property in terms of site risk (chapter 4). Finally, we outline legal controls (chapter 5), the range of management alternatives (chapter 6), and additional information sources (appendixes A and B). If we choose, we can reduce the risks we take with respect to coastal hazards.

Settlement on the lakeshore: a historical perspective

According to Professor D. R. Bush, Case Western Reserve University, archaeological findings document the existence of several different Native American tribes that lived around Lake Erie about 400 years ago. In general, the Iroquois group occupied the lion's share of the area (western, northern, and eastern) and the Algonquin group occupied the central southern area. A 1656 French map bears the name of the lake, "Erie, du Chat," after the Cat Nation as some of the Erie natives were known.

Unfortunately, there is no record of how these Native Americans viewed and responded to Lake Erie coastal processes. Because of the nature of their lives, it is likely that shore erosion and flooding were processes that were lived with and accepted as expected events.

Settlement of the lakeshore by the white man did not begin until more than 100 years after the French explorer, Louis Jolliet, paddled along the north shore of the lake with an Iroquois guide in 1669 (reference 1, appendix B). Explorers like Jolliet undoubtedly observed denuded bluffs and slopes indicating shore erosion, but erosion was of no consequence until the settlers began to build along the lakeshore. The early visitors also observed miles of beaches that cannot be found today.

Notable settlement began after the British were defeated in 1794 at the battle of Fallen Timbers near Toledo, and after they surrendered Detroit in 1796, allowing the peaceful immigration of settlers from the east. Moses Cleaveland, who traveled from Connecticut to Cleveland, Ohio, much of the way along the lake's beach from Buffalo, was one of the most prominent settlers.

Development of the shore took place first near harbors such as Buffalo, Erie, and Cleveland, and then spread to the adjacent shore land. It wasn't long thereafter that shore erosion began to be considered as something more than an acceptable natural process, for now the erosion was impacting upon land that had been surveyed and titled.

Measurements of erosion were first reported in 1838 by Charles Whittlesey, one of Ohio's first geologists, (p. 53, reference 6, appendix B), who wrote:

> When the first settlers of the Western Reserve came along the Ohio shore in 1796, the sandy beach of the lake was occupied as a road throughout, and was used for that purpose east of Cleveland many years.
>
> At the present time, the encroachment upon surveyed lots between the Cuyahoga and Chagrin rivers, is from 10 to 20 rods (a rod is 16.5 feet).
>
> If we except a short distance along the shore west of Conneaut harbor, about 5 miles next westerly of Fairport, and 20 miles rock coast between Cuyahoga and Black Rivers, the entire shore from the State line to the lime rock near Huron, has lost an average of 8 rods in width. The immediate bank is composed of loose earthy materials incapable of resisting the action of the wave, with the above exceptions.

In those days, the lakeshore was still in a natural state, but erosion was a recognized process. It is an important commentary on modern American life that some lakefront property owners in the 1970s and 1980s express surprise at learning that the shoreline is eroding.

The early settlers built up towns such as Buffalo, Cleveland, and Sandusky adjacent to the major streams along the lakeshore. One of the first orders of business was the construction of jetties at the stream mouths to facilitate the burgeoning lake traffic. These structures had the almost immediate effect of trapping voluminous quantities of sand on one side of the structure, thus accelerating erosion and the loss of sand on the other side. In all likelihood, these results were anticipated by the people who built the jetties, but there were no apparent problems simply because there were few people living along the lakeshore in the early 1800s.

Although the number of coastal settlements was limited in the 19th century, the few residents left chronicles of storms and their impact on human developments. For example, E. L. Moseley, a naturalist and teacher, wrote in 1904 of lake levels and erosion between 1857 and 1862 (p. 2, reference 7, appendix B):

The greatest storms of the past century or those which were most effective because occurring at time of highest water were those of 1857–1862.

The water covered the land where the Sandusky Tool Factory stands and the street adjacent so that the workmen went to the building, then a sawmill, in rowboats. It flooded the cellars on the south side of Railroad Street. The part of the city near the end of First Street and east of it was under water. These storms damaged the bridge across Sandusky Bay and the railroad near Port Clinton. Along miles and miles of shore and over hundreds of acres of lowland they killed trees that had stood for centuries. They cut away large slices of Eagle Island at the head of the bay. They cut through the land west of Port Clinton giving an outlet for the Portage River about one-fourth mile farther west than before, but the breach was afterwards closed by the L.S. & M.S.R.R. Co. They built up on the northeast shore of Cedar Point long sand ridges twelve feet high on which hundreds of cottonwoods have since grown to a height of fifty or sixty feet.

The preceding statements may suffice to illustrate the sort of changes effected by northeast gales but those who have seen Lake Erie only when it is calm or stirred by winds of moderate force will be further impressed with its power by a brief notice of particular storms which are remembered by old residents or noted in the journal of the weather observer.

The northeast storms of 1857–1862 are said to have been more frequent and usually of longer duration than those of late years. Regarding this point a number of old residents agree and they are probably not mistaken, for the records of rainfall at the stations in this part of the country where records were kept so early show that the precipitation of 1857 and 1858 has not been equalled since.

By the early 1900s, the lakeshore was much more developed. With development came erosion problems for some and flood problems for others. Instead of moving away from the lake, the property owners chose to stay and fight the lake by building groins, seawalls, and dikes in an effort to stabilize the shore, i.e., to hold it in a fixed position. These structures for the most part succeeded in protecting the shore behind them for a few to a few tens of years, depending upon how well they were designed and built. Eventually, though, the structures were not maintained, were undercut or outflanked, and erosion of the shore resumed until a new generation of structures was built, usually behind the older damaged structures. The retreat has been disorderly.

The pattern of several generations of structures can be observed along many of the older, first-settled lakeshore stretches where remnants of former structures lie submerged off the present shoreline. Yet development on the shore continues. More structures, often better designed and better built, front the shore; but erosion and flooding are still problems.

Many lakeshore owners do not understand the lake or its processes. And with time, those who learn are replaced by a new generation of owners who must learn anew.

One aspect of the problem is reflected by this observation in the International Joint Commission's Lake Erie Water Level Study (p. 223, reference 101, appendix B): "There is a lack of clear understanding by some of the public of the various natural and man-made factors affecting the Great Lakes levels and the reasons for the extreme high and low water levels."

This statement ties into one of the principal purposes of this short book, that is, to enable the lay person to understand the physical processes and natural consequences of living along the lakeshore so that intelligent decisions can be made with respect to buying, selling, building, moving, and finally living with the lake as a good neighbor.

Origin of the lake: a geological perspective

Lake Erie cannot be truly understood without a geological perspective. This view is necessary because of the overriding influence that the geologic deposits and processes have had and continue to have on the evolution of the lake. These deposits begin with the most ancient, the bedrock, which often is overlain by sediment derived from the Pleistocene (Ice Age) glaciers of thousands of years ago, which in turn often is overlain with "modern" deposits that are accumulating at the present along the lake's margin and bottom.

The bedrock consists largely of limestones and dolostones (carbonates), and shales (fig. 1.8). The limestones and dolostones, the gray to cream-colored fossiliferous rocks, make up the headlands and the islands at the west end of the lake, such as at Marblehead and Pelee Island. These rocks also outcrop along the eastern Canadian shore between Port Dover and Fort Erie.

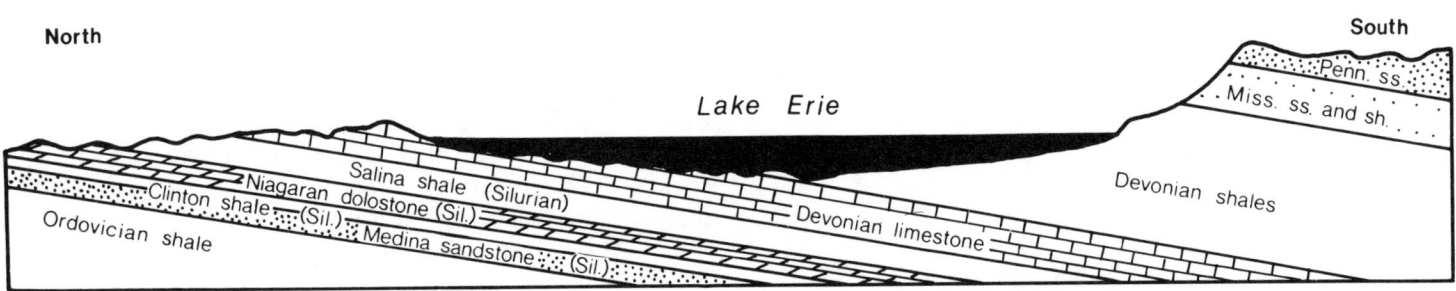

North

South

Lake Erie

Penn. ss.

Miss. ss. and sh.

Salina shale (Silurian)

Niagaran dolostone (Sil.)

Clinton shale (Sil.)

Ordovician shale

Medina sandstone (Sil.)

Devonian limestone

Devonian shales

1.8 Geologic cross section of the eastern end of the Lake Erie basin. The lake occupies the central part of the basin in the less resistant Devonian shale. The more resistant limestones and dolostones to the north and east, and sandstones to the south form the margins of the basin. ("Miss. ss. and sh." is the geologist's abbreviation for "Mississippian sandstone and shale"; "Penns. ss." for "Pennsylvanian sandstone." Modified from *The Geologic Interpretation of Scenic Features in Ohio*, by J. Ernest Carman (reference 20, appendix B).

The shales, the dark gray to black, thinly layered rocks, occur here and there in the banks and bluffs along the south side of the lake from Vermilion to Conneaut. However, from Conneaut to Buffalo, the shale outcrops almost continuously along the shore with the elevation of the shale surface rising from west to east. The particles making up these sedimentary rocks were deposited in Silurian and Devonian times between about 420 and 320 mil-

lion years ago. The carbonates formed in shallow, tropical seas, and the shales in deeper, marine basins.

The rocks were deformed later by mountain building activity in the Appalachians, and they now slope to the southeast at about 20 feet per mile over most of the Lake Erie region. Any younger overlying rocks that covered them were subsequently stripped away by erosion. Differences in the erodibility of the softer shale and more resistant carbonate rocks led to the development of an irregular landscape. Evidence of this difference in erodibility can be seen at both ends of the lake. At the western end, the carbonates floor the shallow western basin and make up the more erosion-resistant Lake Erie islands. At the eastern end, the carbonates form the Niagara escarpment—the reason for Niagara Falls. In between lie the more easily eroded shales that now floor the Lake Erie basin. Whether the present-day Great Lakes axes represent major preglacial river valleys (fig. 1.9) is unclear, but it is quite likely that the paths of the Pleistocene

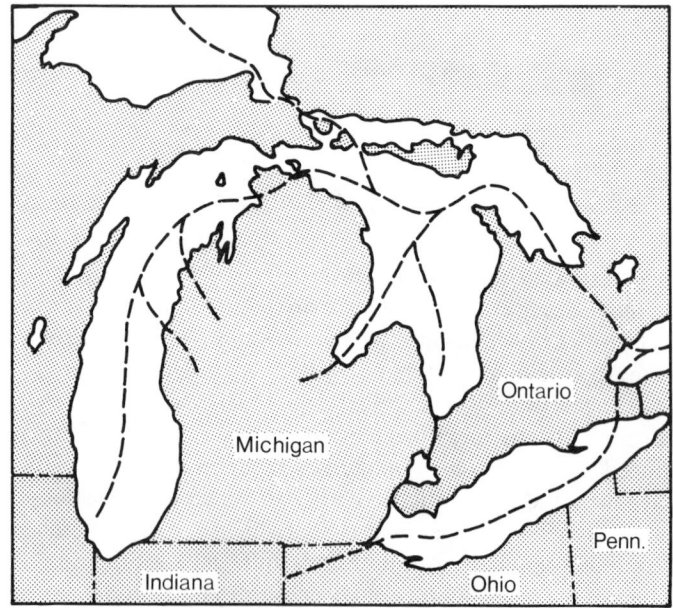

1.9   Probable preglacial valley system in the Great Lakes region. Path of the valley floors. Modified from *Geology of the Great Lakes*, by Jack L. Hough (reference 17, appendix B).

glaciers followed depressions cut into these sedimentary rocks by ancient rivers (reference 17, appendix B).

The rather commonplace sedimentary bedrock, aside from having a significant effect on the location and physical dimensions of the lake, is extremely valuable in both economic and scientific ways. Silurian salt associated with the carbonates is presently mined under Fairport Harbor and Cleveland, and formerly from beneath Detroit. Gypsum, also associated with the carbonates, is mined near Port Clinton as are the carbonates from Marblehead and Kelleys Island that are used for riprap (armor stone). In addition, unique geologic specimens include the spectacular crystals of celestite (a strontium mineral) in the caves at South Bass Island, the well-preserved Devonian corals in the Columbus limestone, the enigmatic concretions found in certain zones of the Devonian Ohio shale, and the outstanding Devonian fossil fish also found in the Ohio shale. Excellent displays of the latter may be seen at the Cleveland Museum of Natural History.

Capping much of the bedrock over the Lake Erie basin are deposits laid down by or associated with the younger Pleistocene glaciers, deposits from the Great Ice Age of a few tens of thousands of years ago. In this material are found igneous and metamorphic rocks that are totally alien to the underlying Lake Erie basin's sedimentary bedrock. These so-called erratics are from outcrops in Canada where the enormously powerful glacial ice sheets moved south across the underlying rocks, eroding and transporting some of them for hundreds of miles. In addition to the erratics, the glacial debris includes the carbonates and shales from the Lake Erie basin, indicating more local derivation. Moreover, some of the bedrock surfaces exhibit polished, scratched, and gouged surfaces that reflect the scouring action of the rock-laden ice moving slowly across the bedrock surface. Excellent examples of such features may be seen at Kelleys Island State Park.

Geologists know from different lines of evidence that the glaciers moved southward from the Canadian ice caps. Once the glaciers reached the Great Lakes region it is likely that they flowed along the preexisting valleys already cut into the bedrock by rivers (fig. 1.9). Specific evidence from the Lake Erie basin, such as the northeast-southwest orientation of glacial striations, and the occurrence of erratics at the west end of the basin derived from bedrock at the east end of the basin, indicates that the glaciers traversed the Lake Erie basin from the northeast to the southwest along the depression cut into the Devonian shale that makes up much of the lake's bedrock. The advancing ice may have deepened and widened the basin in which Lake Erie formed. Such enlargement is suggested by the closed depression shape of the eastern lake basin where present water depths exceed 200 feet.

Although there were four major ice advances (glacial periods) in the Pleistocene epoch, only the last advance, known as the Wisconsinan, is clearly represented in the sedimentary deposits of the Lake Erie basin. These deposits reflect the diversity of processes associated with the glaciers. They range from till

(mixed debris left in place by the melting glacier with little or no transport by water), to outwash (sand and gravel transported from the glacier margins by streams), to lake deposits (sediment transported from the glacier into lakes, with clay and silt deposited in the deepest parts of the lake, and sand and gravel laid down along the former lake margins).

Wisconsinan lake deposits are of special significance to our understanding of the present lake's behavior. These lake deposits border Lake Erie's southern margin, and were deposited by higher lakes ancestral to the present-day Lake Erie. The rise and fall of these early lakes reflect the dynamic history and nature of the Great Lakes in the not-too-distant past.

About 14,500 years ago a major glacial lake, Lake Maumee, formed from the meltwater at the margin of the glacial ice. The lake ultimately occupied much of what is now Lake Erie, as well as a considerable area beyond the present lake's shores. Although the climate was much cooler, this early lake had a shoreline along which the waves and wave-caused currents worked to erode and transport sediment—similar to what is happening in present-day Lake Erie. The sand and gravel was transported along the shore to form the beach ridges, and the finer particles, the clay and silt, were transported offshore to settle in the quiet, deep lake waters. As the glacier retreated and as new lake outlets were uncovered, a series of glacial lakes formed at successively lower elevations (fig. 1.10). Each lake level formed a beach, a ridge of sand left behind as tangible evidence of former shoreline occupation (table 1.1). These features are the beach ridges used by

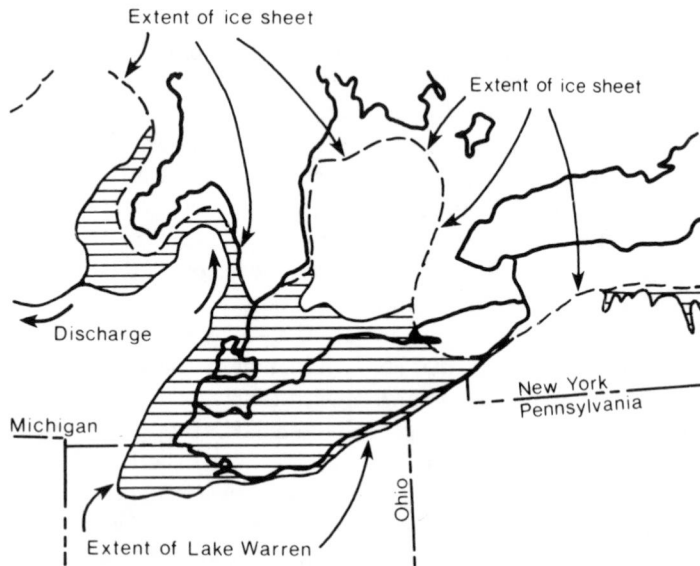

1.10   Map of Lake Warren (686 foot elevation), one of the ancestral lakes (see table 1.1). Note that this lake covered present-day areas of Windsor, Detroit, and Toledo, and discharged into the Huron basin. Modified from *Geology of the Great Lakes*, by Jack L. Hough (reference 17, appendix B).

Table 1.1   Lake Erie's ancestral glacial lakes. The elevations of these ancient lakes are considerably above Lake Erie's present average level of 570.4 ft. From table 7 in *Geology of the Great Lakes*, by Jack L. Hough (reference 17, appendix B).

Glacial lakes	Elevation (feet)	Age (years before present)	Outlet
Maumee Ia	780	14,500	Wabash River, Indiana
Maumee Ib	774	14,450	Wabash River, Indiana
Maumee II	764	14,250	Wabash River, Indiana
Maumee III	753	14,100	Wabash River, Indiana
Whittlesey I	740	13,950	Grand River, Michigan
Whittlesey II	732	13,850	Grand River, Michigan
Arkona I	711	13,500	Grand River, Michigan
Arkona II	700	13,350	Grand River, Michigan
Arkona III	695	13,300	Grand River, Michigan
Warren I	686	13,050	Grand River, Michigan
Warren II	680	13,000	Grand River, Michigan
Warren III	670	12,900	Grand River, Michigan
Wayne	660	12,850	Mohawk River, New York
Grassmere	640	12,800	Mohawk River, New York
Lundy	620	12,750	Mohawk River, New York

the Native Americans and early settlers as paths, campsites, and burial grounds. The intervening areas, covered by glacial silt and clay that settled out in the deeper portion of the ancestral lakes, have become rich farmland. These old lake plain areas are nearly flat and, together with the associated ridges, surround the present lake.

Lake levels continued to drop for 2,000 years as the last Wisconsinan ice melted in response to the increase in the earth's surface temperature (table 1.1). However, table 1.1 does not show the lowest level of early Lake Erie as determined by Mike Lewis, a Canadian geologist. He found remains of shallow water plants at the bottom of Lake Erie, about 130 feet below the present level of the lake. Geologists infer from this and other evidence that as the glacier melted away from the Fort Erie-Buffalo area, glacial Lake Lundy, formerly dammed by the glacier, flowed as a torrent eastward into the area presently occupied by Lake Ontario. This draining caused a rapid drop in Lake Lundy's level to about 130 feet below the lake's present level. Lake Lundy was lowered to this level because the Fort Erie-Buffalo area had been depressed by the great weight of the ice (isostatic deformation). When the ice disappeared from the eastern end of the basin, the water was no longer held back and flowed across the topographically low area.

Later, rebound of the Fort Erie-Buffalo area raised the land on the east end of the basin, which again blocked the flow of water. This damming effect led to the final filling of the basin and the formation of the lake at its present level about 2,000 years ago.

The filling of early Lake Erie was not without change in the surrounding land. As the climate warmed, vegetation took hold; first the cooler climate conifers, and later the present-day deciduous plant varieties. Tributary streams had eroded their valleys down to the earlier lower lake level. As lake level rose to the present elevation, the mouths of these stream channels and valleys were flooded (drowned).

The more recent geologic history of Lake Erie is the continued sculpting of the shore by waves and wave-caused currents, eroding much of the shoreline, and sometimes depositing sediment, forming the sandy spits and headlands such as Presque Isle, Long Point, Rondeau, and Point Pelee. The overall perspective is one of constant change—of spectacular shifts in shoreline and lake level, and of modification of topography and sediment by waves and currents. These continue to operate today, and because of their effect on erosion and flooding they are of immediate concern to the coastal dweller and property owner.

2 Coastal dynamics

Lake Erie's coastal zone is in perpetual motion. At the shore, conditions range from turbulent to tranquil in the never-ending struggle between the waves and the land (fig. 2.1). The zone, in physical terms, is exceedingly complex. A knowledge of major variables such as wave height and direction, long-shore drift, lake level, and the nature of shore deposits is crucial to understanding the system. These variables are already complex, but one also must consider the effects of man-made jetties and the shore protection structures.

From the coastal dweller's and property owner's perspective an appreciation and some knowledge of these variables is most important in learning to live with the shore. If you are going to build shore protection structures, what can you do to avoid having the structures destroyed or damaged during storms? How can you avoid causing damage to adjacent beaches or property? To answer such questions one should initially consider the two major variables: changing lake levels and waves.

Lake level changes: as predictable as the weather

Lake level is of paramount importance to the lakeshore property owner and resident because the times of greatest erosion and flooding are during the highest lake levels. The reasons for this

2.1 Contrasts in lake conditions near Huron, Ohio. Calm conditions on a November day (A) and a stormy April day (B).

are simple. At times of high lake level, waves (the cause of shore erosion) are able to break closer to the shore, and when waves reach the shore there is erosion. Similarly, the higher the lake, the closer the lake is to flood stage for the low-lying communities along the shore. In the mid-1960s when lake level was quite low, shore erosion and flooding were not regarded as problems. During low lake levels, beaches are at their widest and absorb the energy of the waves, keeping the waves from causing significant erosion. In addition, low lake levels are not likely to be elevated by storm winds to sufficient heights to cause flooding.

In contrast, during times of high lake level, as in the early 1970s and again in the mid-1980s, shore erosion and flooding are regarded as major problems. Storm waves break close to or directly on the shore because beaches are largely submerged. The lake is high enough to allow storm winds to cause flooding of low-lying areas. Such flooding occurred along the eastern New York shore in the fall of 1985.

In addition to shore erosion and flooding, changing lake levels affect a wide range of activities including the generation of hydroelectric power, the amount of cargo transported by lake freighters, the area of marshland available to waterfowl, and the area of beach available to summer vacationers. Because of the significant effect of changing lake levels there were two major International Joint Commission (IJC) studies in a 15-year period that dealt with lake level problems. The first, completed in 1973, was initiated because of the low lake levels of the 1960s. The second, completed in 1981, placed special emphasis on Lake Erie

and was initiated because of the high lake levels of the early 1970s (references 100 and 101, appendix B). Most recently in the mid-1980s record-setting high lake levels have generated a new hue and cry from besieged shore owners for assistance and relief from the problems of flooding and shore erosion.

Changes in lake level, fluctuations if you will, have become much more marked since the 1920s. The effects of the fluctuations are regarded as more serious because of the great increase in shoreland development—more people are affected. Nevertheless, the most basic question is: what do we really know about lake level fluctuations? (What causes them? Can they be predicted?)

Lake level is commonly measured relative to a reference level, which is referred to as a *datum*. Lake Erie's reference water level is the International Great Lakes Datum (IGLD), which was established by Canada and the United States in 1955. This datum is 568.6 feet above mean sea level at Father Point, Quebec (near the mouth of the Saint Lawrence River). The datum usually is referred to as *chart datum* or *low-water datum*, and is about two feet below Lake Erie's long-term average level of 570.4 feet. In discussing the level of Lake Erie one would say, "the lake is about three feet above low-water datum," meaning that Lake Erie's elevation is about 571.6 feet. The lake's level is measured with water level gauges that are located at most of the major harbors along the lake shore. The instruments that measure the lake level are located in wells, which damp out the effect of waves so that the measurements are that of a nearly horizontal surface.

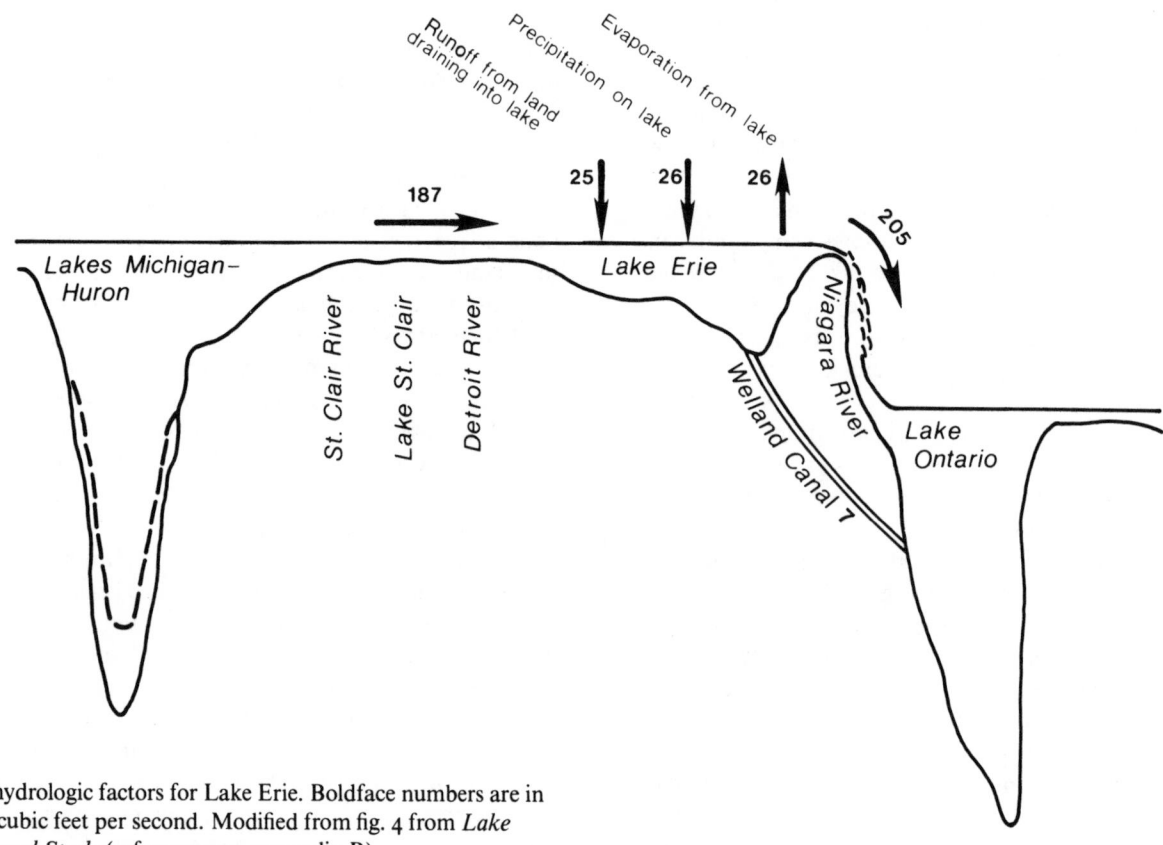

2.2   Major hydrologic factors for Lake Erie. Boldface numbers are in
thousands of cubic feet per second. Modified from fig. 4 from *Lake
Erie Water Level Study* (reference 101, appendix B).

The level of the lake reflects the amount of water stored in the lake at a given time. Hydrologists characterize the variation in the amount of storage as the *water budget*. This budget is simply water added minus water removed. Sources of addition are precipitation over the lake, inflow from the upper Great Lakes via Lake Saint Clair and the Detroit River, surface runoff from the streams of the Lake Erie drainage basin (e.g., Ohio's Maumee River and Ontario's Grand River), groundwater that seeps into the lake, and man-made water diversions into the lake.

Water is removed from the lake by outflow down the Niagara River, evaporation from the lake surface, groundwater seeping out of the lake, and man-made diversions out of the lake (e.g., the Welland Canal and municipal and industrial water intakes).

Hydrologic factors of water gain or loss are measurable (fig. 2.2), and excesses one way or the other cause lake level to rise or fall. Figure 2.2 shows average values where water gain is in balance with water loss, but the reality of nature is short-term imbalance. With the exception of poorly understood groundwater flow into and out of the lake, good records exist for the dominant controlling factors (i.e., precipitation, runoff, evaporation, inflow, outflow, and diversions); these allow investigators to determine the causes of lake level fluctuations.

Both seasonal and longer-term variations in lake level are largely the effect of precipitation—the amount of rain, sleet, and snow that falls over the lake and its drainage basin and associated runoff. For example, the seasonal high in June and July is due to relatively greater amounts of spring rainfall and river

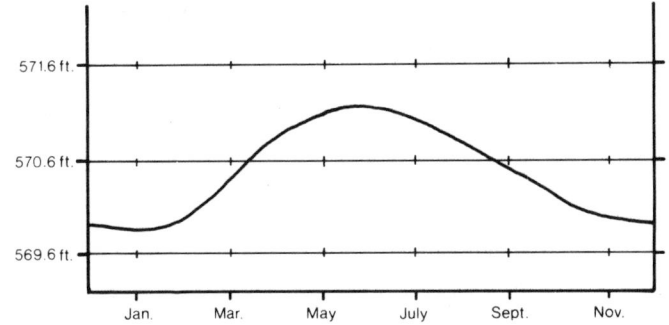

2.3 The average annual lake level elevation curve for Lake Erie, 1900–1984. Modified from *Coastal Geomorphology and Geology of the Ohio Shore of Lake Erie*, by C. H. Carter, D. E. Guy, Jr., and J. A. Fuller (reference 21, appendix B).

input, and relatively smaller amounts of evaporation. In contrast, the seasonal low in December and January follows the relatively greater amount of summer evaporation and relatively smaller amounts of precipitation and stream runoff in the fall (fig. 2.3).

Note that there is a lag time, both seasonally and for the longer term, between the precipitation peaks and the lake level peaks (figs. 2.3 and 2.4). Times of higher rainfall cause increased stream flow into the lake and add to the groundwater input. Such times are usually cooler with more cloud cover so that less water is removed by evaporation. The upper lakes experience the same wet weather, and inflow from Lake Huron and Lake St. Clair is increased. The lag time reflects the time it takes for these water inputs to reach Lake Erie. Clearly, wet seasons or a string of wet years produce higher lake levels. Conversely, times of reduced precipitation reduce stream input, lower the inflow from the upper lakes, and probably reduce groundwater input. Such times tend to be warmer with less cloud cover so that evaporation is increased. These processes work together to lower lake level.

The close correspondence between recorded precipitation and lake level fluctuation has been documented by Dr. Frank Quinn of the National Oceanic and Atmospheric Administration's Great Lakes Environmental Research Laboratory (fig. 2.4). Although the correlation is not perfect, years of below-normal precipitation such as in the 1930s and 1960s were also times of low lake level. When several years of above-normal precipitation occur, as in the early 1950s, early 1970s, and again in the middle 1980s, lake level is high. The maximum range in Lake Erie levels is 4.7 feet, which is the difference between the record 1934 low (568.0 feet) and the record 1973 high (572.7 feet). The lake's large storage capacity and the small inflow and outflow channels prevent larger or more rapid changes in lake levels.

Human impact on lake level

Two important questions are: What effect has man had on Lake Erie's level, and can we control lake level? The major man-made effects are those related to the diversions. Water is diverted into Lake Superior from Canada, diverted out of Lake Michigan at Chicago, and diverted out of Lake Erie at the Welland Canal. The net effect of these diversions is to lower the level of Lake Erie by about three inches from what it would be in a totally natural state (reference 100, appendix B). Although rumors abound concerning control structures in the Niagara River designed to keep the lake artificially high for commercial navigation, no engineering or scientific evidence confirms such rumors. The level of Lake Erie is natural except for man-made diversions that have lowered its natural level by about three inches. Thus, natural factors are the primary controls on lake level. Artificial diversions could be altered (i.e., increased or decreased), but such actions often create new problems.

Storm impact on lake level

In contrast to the relatively slow, seasonal, and longer-term lake level changes, wind storms generated by high and low pressure systems that track north and south of the lake can cause major, rapid, short-term changes in lake level. These fluctuations are known as *storm surges*, and they raise or lower the level of the lake, particularly at its confined eastern and western ends. In essence, the interaction of storm winds with the lake surface (*wind stress*) leads to transport of large volumes of lake water in the direction that the wind is blowing. Once this water reaches the lakeshore it continues to rise (*storm surge*) until the water

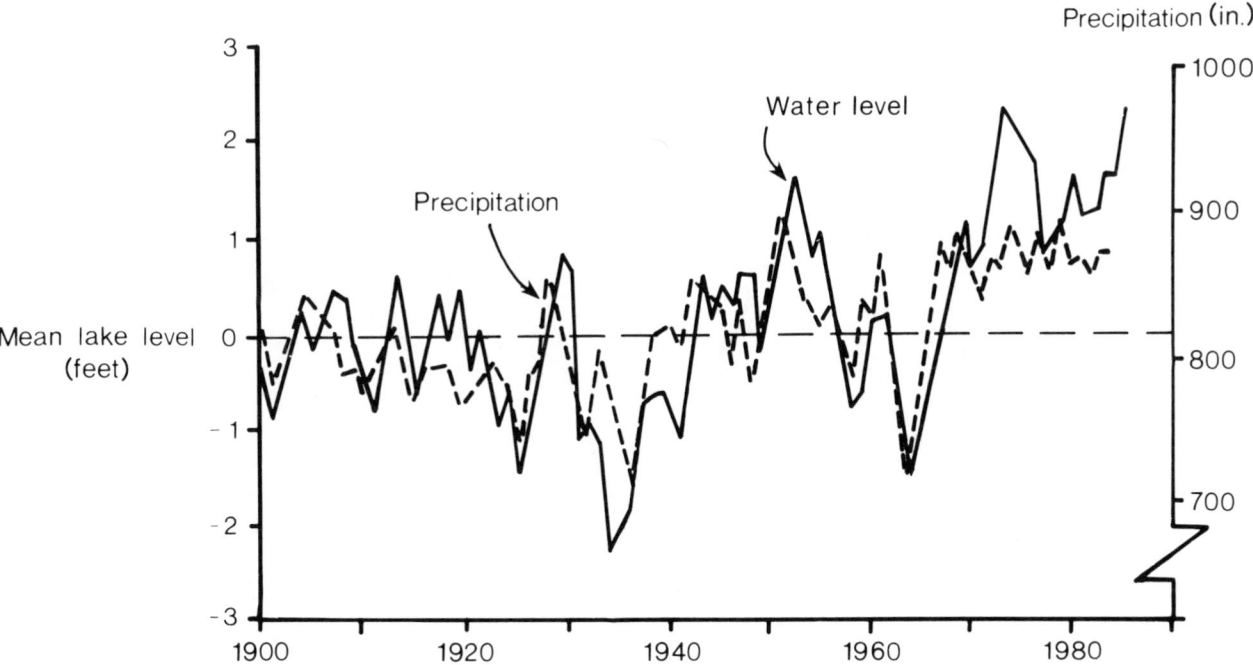

2.4   Lake level and precipitation curves for Lake Erie, 1900–1984.
Note the fairly close correspondence between precipitation (dashed
line) and water level (solid line). Modified from a figure presented by
Dr. Frank Quinn at the Great Lakes Levels Briefing, International
Joint Commission, Washington, D.C., 1985.

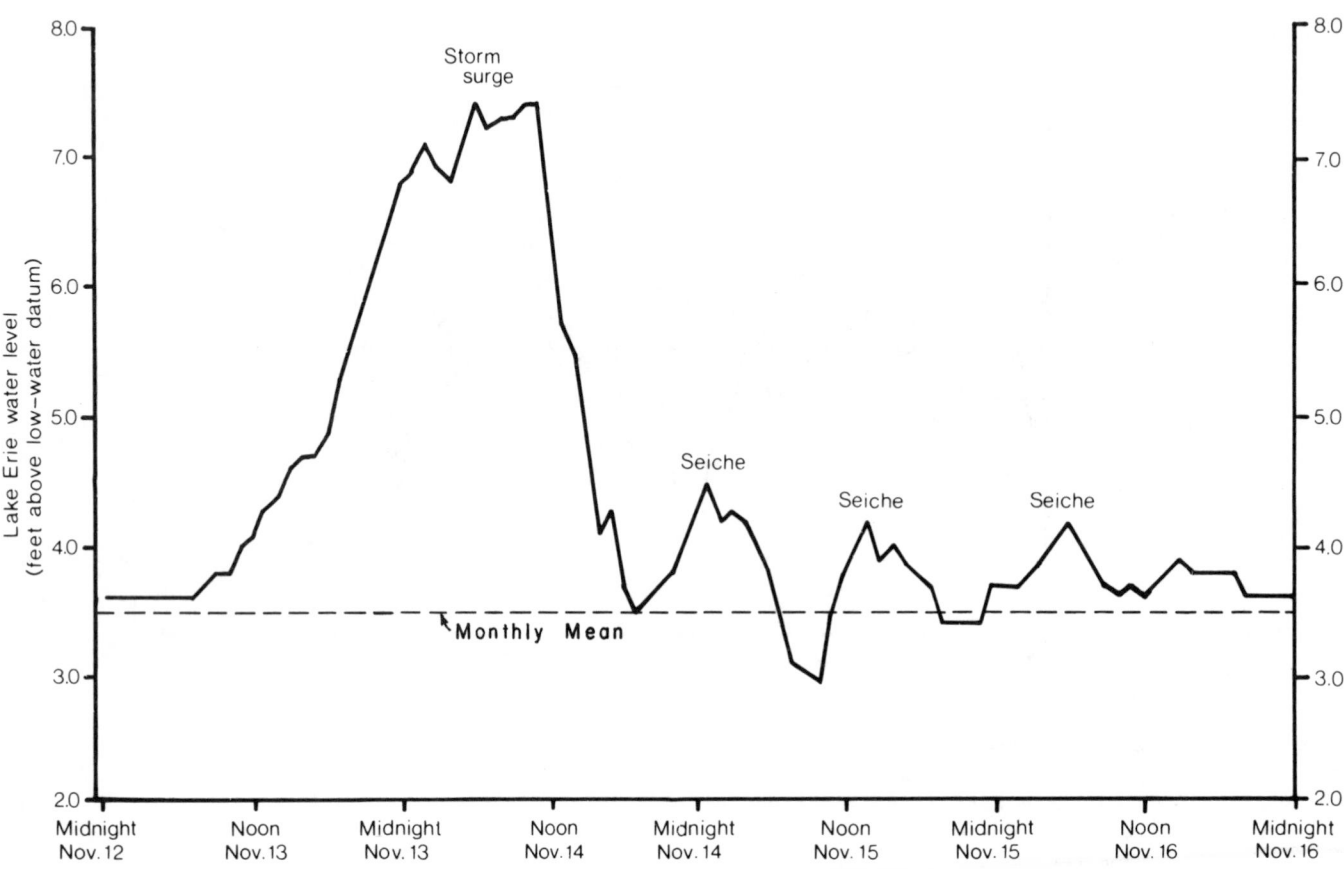

level is high enough to cause water on the lake bottom to flow in the opposite direction. This reverse motion, which may continue for two to three days, is characterized by back-and-forth oscillations of water level in the entire lake. This effect is known as a *seiche*, and is comparable to water sloshing back and forth between the ends of a bathtub. In this case, the shallow Lake Erie basin is analogous to a tub. The initial and greatest rise in lake level (the storm surge) causes the most damage along the shore because of the marked rise in water level and the accompanying large wind waves. After the initial surge, smaller seiche high-water levels follow (fig. 2.5).

Two of the historically biggest storms in terms of storm surges struck Lake Erie in the 1970s. The first, in November 1972 (fig. 2.5), was considered a "100-year storm." Northeast winds of up to 70 mph were generated by a low-pressure center that passed south of the lake. The winds produced deep-water waves of at least twelve feet in height and a wind-caused rise in lake level (storm surge) of about four feet at Toledo (references 43 and 44, appendix B). The second storm, in April 1979, also caused an extreme surge, but this time at the eastern end of the lake. Southwest winds up to 114 mph were generated by a low pres-

2.5 Lake level curve at Toledo, Ohio for November 13–14, 1972, storm. Note prominent storm surge peak followed by three seiche peaks. Modified from *The November 1972 Storm on Lake Erie*, by Charles H. Carter (reference 44, appendix B).

sure center that moved along a west-to-east path, north of the lake. The winds produced deep-water waves of at least sixteen feet in height, and a lake level rise of about eight feet at Buffalo (reference 45, appendix B). The surges for similar storms at the east end of the lake are greater than those at the west end of the lake because the higher shore elevations at the east end keeps the water confined to the lake. The much lower shore elevation (generally less than five feet) at the west end of the lake allows the lake water to flood inland, limiting the amount of lake level rise.

The weather, therefore, can cause lake level fluctuations that range from the short term (the storm) to the intermediate and long term (the up or down trends of several years). With over 120 years of water level records collected by the U.S. National Oceanographic and Atmospheric Administration (formerly the U.S. Lake Survey), we can look for patterns to help us predict when the lake is going to be high or low. The short-term fluctuations, however, are related to storms and storm surges and remain as unpredictable as the weather. Nevertheless, well-defined weather systems, particularly low-pressure centers, are known to generate hazardous conditions as they move into the Great Lakes region. At such times the weather services provide forecasts of lakeshore areas likely to be affected by flooding and shore erosion. Residents in threatened areas (low relief, easily eroded shores) may have from several to only a few hours of advance warning to take necessary precautions, as the storm surges build to their maximum set-up.

The seasonal fluctuations are much better known. For example, the Detroit district of the U.S. Army Corps of Engineers publishes a *Monthly Bulletin of Lake Levels for the Great Lakes* in which six-month forecasts are made for the probable levels and range of levels for the Great Lakes (reference 40, appendix B). In general, their predictions are quite good, but not perfect because of unexpected weather patterns. The fall of 1985 was a good example. In May 1985 the predicted Lake Erie level for November 1985 was 572.0 feet with a probable range of 571.6 to 572.4 feet. The recorded November average level was 572.6 feet, 0.6 feet above the predicted level. What was the reason for the marked difference between the forecast level and the actual recorded level? The weather! Abnormally heavy rainfall over the Great Lakes region (rainfall in some months was two to three times the average monthly amounts in the lake basins) and an associated decrease in evaporation caused the unpredicted and statistically unlikely record-high monthly levels of November 1985.

Those who have worked on the lakes for a number of years intuitively expected Lake Erie's record levels of 1972–1973 to drop at least to the long-term average level of about 570.4 feet by the end of the 1970s. Such a drop had occurred following high lake level years since 1860 (fig. 2.6). However, this pattern was not repeated. Lake level stayed at least 0.7 feet higher than the long-term average level for 14 years, a new record for sustained high levels on the lake. This 14-year period also included record-setting levels for each month either in 1972, 1973, 1985, or 1986;

statistically, this is a quite improbable situation. What is happening? Again, the records indicate that the weather, or more properly the climate, has changed over the past 125 years. In recent years, greater than normal precipitation for longer periods has caused the prolonged high lake levels (fig. 2.4).

Geologists Barry Cohn and Joseph Robinson (reference 42, appendix B) have calculated controlling factors on lake level changes and determined cycles of 8, 11, 22, and 36 years' duration. On this basis they predicted a 1985 high level for Lake Erie of about 570.5 feet, which turned out to be about two feet lower than the recorded high for 1985. They predict a return to high lake level in 1993 at approximately 571.2 feet. If this prediction is also two feet too low, the lake will set a new high-water record in the 1990s!

The point is not the accuracy of such predictions, but rather that lake level fluctuations will certainly continue over the short to intermediate term. And shore erosion is the rule during times of high water, not an unusual exception to the rule. To add to the uncertainty of the shoreline's future, longer-term geologic cycles measured in hundreds to thousands of years, and for which there is no historical record, could produce additional unexpected fluctuations.

Waves: the sculptors of the shoreline

Waves, like lake level, are of paramount importance to the lakeshore property owner and resident. Wave energy causes shore

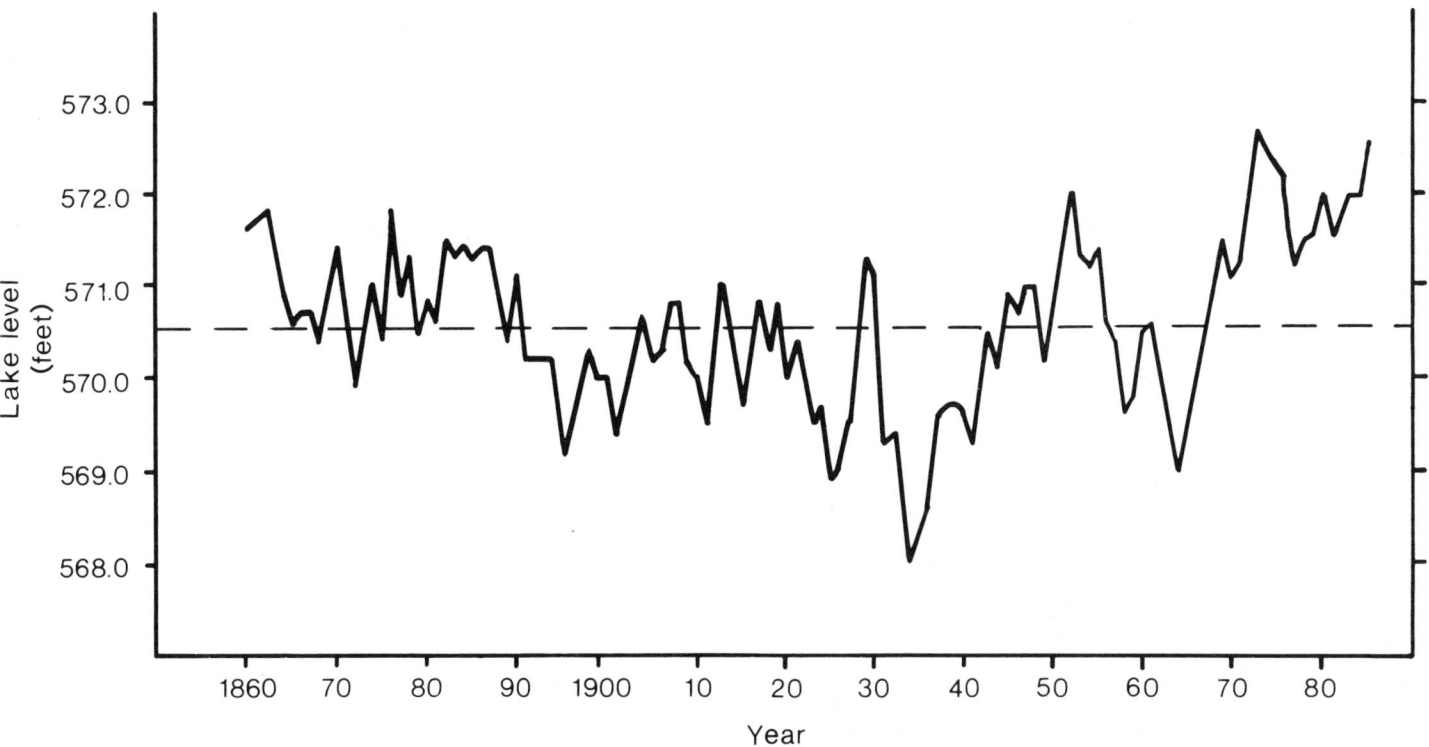

2.6 Lake Erie's water-level curve, 1860–1985. Dashed line indicates the long-term average lake level. Data from National Oceanic and Atmospheric Administration, National Ocean Survey.

erosion; and waves, inseparably associated with storm surges, contribute to much of the damage related to coastal flooding (fig. 2.7).

The interaction of the wind with the lake surface produces waves. The everchanging, fascinating appearance of the lake is due to the variability in speed and direction of the winds that blow across Lake Erie. This variability creates a myriad of scenes on the lake surface.

Of the different characteristics of waves, wave height is most crucial to coastal processes. The amount of energy, and thus the potential for doing work such as eroding the shore, is largely related to wave height. Most scientists consider wave energy to

2.7 Storm waves at Port Maitland, Ontario.

be proportional to the square of wave height. Waves can range from nearly imperceptible ripples to storm-driven white horses, and their size is largely dependent on three factors: wind speed, wind duration (the length of time that the wind blows from a given direction), and *fetch* (the over-water distance across which the wind blows). The stronger the wind, the longer it blows, and the greater the fetch, the larger the waves. Estimates of potential wave heights for Toledo, Cleveland, Erie, and Buffalo (reference 46, appendix B) reflect these factors. For example, Buffalo, at the east end of the lake, is exposed along the lakeshore only to winds from the west-southwest, and only waves from the west-southwest reach the Buffalo shore. On the other hand, the Monroe, Michigan, shore at the west end of the lake is exposed only to winds and waves from the northeast and southeast. Therefore, the same storm may have radically different effects on different stretches of shoreline.

The prevailing winds for Lake Erie are mid-latitude westerlies that frequently blow along the long axis of the lake from the west end to the east end. Such winds, especially when strong and blowing for a prolonged period of time, sweep over an appreciable expanse of water (maximum fetch of the lake) producing waves with estimated heights up to at least 16 feet at the east end of the lake. Winds that blow out of the east at high speeds and for long durations can produce waves of similar heights at the west end of the lake, but they are much less frequent. In both cases the results are powerful waves with great potential for destruction. Such conditions produce storm surges, and the combination of waves atop rising water level make the ends of the lake particularly susceptible to storm damage. The easterly windstorms, although infrequent, are the memorable "northeasters" to the lakeshore residents at the west end of the lake in Ontario, Michigan, and Ohio.

In spite of the apparent windiness of the lake, the majority of the time the lake surface at any specific shore location is relatively calm with waves less than 0.5 feet in height. For example, even along the more exposed areas of the shore at Erie, Pennsylvania, and Cleveland, Ohio, the lake surface is characterized by calm conditions, or by waves less than 0.5 feet high about 80 percent of the time (reference 46, appendix B). This apparent calmness is deceptive and may create a false sense of security to one who hasn't experienced the moods of the lake. The calm is broken by the episodic character of natural processes such as flooding and shore erosion, which can take place at infrequent and unpredictable times during storms.

Shore erosion and deposition: energy and materials

When a strong wind blows onshore for an appreciable length of time large waves are generated and move in the direction of the wind. The energy of the wind is transferred to the water. As the waves reach the shallow nearshore waters, they are affected by the bottom and begin to steepen and break. The change in wave form is apparent in the surf zone when waves are running. Although some energy is lost as the waves break, the transformed waves continue to move shoreward. The wave energy may be damped out by a gently sloping bottom, absorbed on the beach, or expended against the back beach and shore slope. When waves reach the shore, their energy is spent in erosion and transport of shore materials. The waves may move beach sand or impact directly against the shore, forcing air or water into cracks and crannies to quarry the shore, or it may carry sand to abrade and sandblast the shore.

The amount of erosion is roughly proportional to the amount of wave energy reaching the shore (wave height), but the resistance of the material making up the shore is an important factor in determining the rate at which erosion takes place. For example, the sand in the dunes at Point Pelee and Presque Isle erodes much more rapidly than the rock in the bluffs at Lakewood, Dunkirk, and Port Maitland. More specifically, erosion rates are usually less than 1 foot per year for rock, 1–2 feet per year for clay, 3–5 feet per year for interlayered clay and sand-silt, and too variable to summarize for sand. Sand is the only sediment that can build lakeward as well as erode landward. Wave action can erode a sandy shore or deposit sand on the shore. The wind may act on such deposits to form dunes and build up the shore.

Erosion rates are also dependent on lake level. The lake level fluctuations noted earlier introduce still another variable into the erosion picture. During a period of rising lake level and high water, the erosion rate increases significantly because the beach narrows and larger waves can reach the shore. During low water the erosion rate decreases because there is a wider zone in which

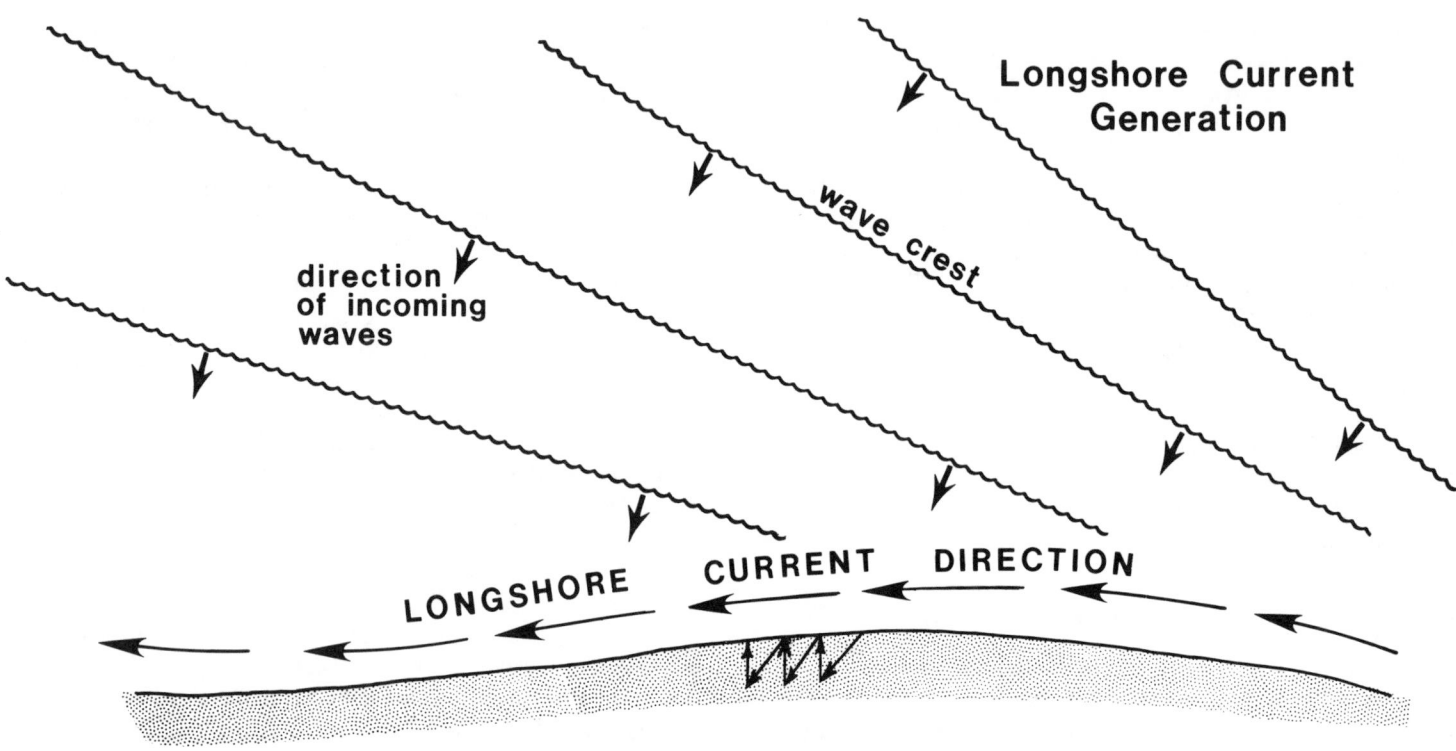

Longshore Current Generation

direction
of incoming
waves

wave crest

LONGSHORE   CURRENT   DIRECTION

2.8 Longshore currents are formed by waves striking the beach at an angle. Sand is transported alongshore by the currents and in a zigzag fashion by wave uprush and backwash.

the waves can expend their energy. If beaches reform or widen, more natural protection is afforded against wave energy.

Not all wave energy goes into erosion. Most waves approach the coast at an angle. As a result, one end of the wave enters shallow water and slows down, allowing the deeper portion to outrun the shallow end. This difference causes the wave to bend or be refracted. Refracting waves may hold water on shore causing a longshore current to develop (fig. 2.8). Longshore currents flow parallel to the shoreline and expend energy in the transport of sediment. Longshore drift of sediment is important to beaches and is discussed later in this chapter.

The removal of material from the lakeshore bank, bluff, or slope, keeps the shoreland slope in a unstable condition. If there were no waves or longshore drift, slope processes would gradually work to produce a stable slope. Under natural conditions, however, the waves erode the toe of the slope causing it to steepen, thereby keeping the slope processes active. At times of low lake level, waves are less likely to reach the slope at the back of the beach. Material accumulates at the toe of the slope so that the overall slope is less steep. When such a slope becomes relatively stable it is said to be at its natural angle of repose. When lake level rises, waves can reach the toe of the slope more easily, causing steepening so that slope processes resume their activity and erosion accelerates. As a rule of thumb, a denuded slope free of vegetation is probably unstable whereas a slope covered with vegetation has not been subject to erosion for some time.

Lakeshore erosion rates represent averages. In fact, erosion does not proceed at a linear (constant) rate. Instead it proceeds fitfully, in a series of stops and starts that are controlled largely by storms during high lake levels. Known erosion rates are generally much higher during high lake level periods because the energy of the waves reaches beyond the beach.

The material making up the shore is also important to the nature of the erosion process. If the shore material has sufficient strength to remain stable even though the slope is steepened or the bluff undercut, then failure (erosion) may not take place until months or even years after the damaging wave erosion.

Lake Erie is dynamic, constantly changing in level and in wave character. The shoreline is constantly trying to come into a stable adjustment with these dynamics. Over the short term, the changes are so frequent that the shoreline is in continual adjustment, i.e., the shore is more likely to be unstable than stable. Change in response to energy, not constancy, is the rule (fig. 2.9).

Ice: friend or foe

Fortunately for the shoreland property owner, ice forms along the lakeshore and covers much of the lake each winter. Ice has a major damping effect upon waves and, by association, storm surges. Ice generally covers most of the lake between mid-January and mid-March. Because the ice acts as a protective barrier between the wind and the water, waves cannot form; thus there is no erosion or flooding from the lake waters. Generally, the shorefast ice that forms adjacent to the shore in December protects the shore from erosion. Beaches freeze, giving additional strength to the shore. The occasional intense wind storms that occur in the winter therefore do not have much effect on the lakeshore. In the spring thaw that usually occurs between mid-March and mid-April, however, when the ice breaks up, large floes can be pushed onto the shore by wind and waves. Such floes often damage or destroy man-made structures.

Gravity: mass wasting of the shore

Shore erosion is not simply due to wave action. Many are the property owners who have built seawalls, groins, and other structures to defeat waves, only to see their property slump, slide, or fall behind the structure. Geologists use the term *mass wasting* to describe such downslope movement of material due to gravity. When a slope becomes steepened or the strength of the material making up the slope becomes weakened or additional load is placed on the material, the force of gravity may become dominant and the slope may fail.

Mass wasting includes several forms of downslope movement, but along the lakeshore the most common types are *falls*, where material simply collapses or topples; *slumps*, where material moves downslope along a curved failure surface; and *debris flows*, where material simply flows down the slope as a chaotic mass. Successive slumps generally produce steplike forms along the slope, whereas debris flows generally produce a slope with an irregular surface.

Wave erosion is the principal mechanism for increasing the steepness of slopes. Processes that decrease the strength of the shore material or add weight to the top of the slope include the infiltration of water from melting snow and rainfall. When combined with the thawing of ice in the shore deposit, the strength of the shore material is reduced to a level where the slope is unstable and mass wasting takes place. This is why mass wasting is so common in the spring. Constructing buildings near the edges of slopes can have a similar effect to water infiltration.

The actual process of mass wasting can involve rapid, almost instantaneous displacement of literally thousands of cubic yards of material, as has occurred, for example, in slumps along the high bluffs of Canada's north shore. The process also includes relatively slow, but persistent displacement of cubic inches of material caused by freeze-thaw cycles on unvegetated lakeshore slopes.

2.9 Changing shore. (A) Winter scene shows protective shorefront ice and lake ice (background). (B) Slope failure produces debris at base of bluff. If this process continued and the debris were not removed, the slope would approach its natural angle of repose. (C) Waves remove debris and attack the toe of the slope, creating a new unstable bluff face. (D) View of unstable bluff during fair weather. Photography provided through the courtesy of the Ohio Geological Survey.

A

B

C

D

A yearly erosional cycle during a high lake level period on Lake Erie generally includes mass wasting in this way: steepening or undercutting of the bluffs or banks by waves in the summer and fall; a static period in the winter caused by freezing temperatures and lake ice; and then mass wasting in the spring caused largely by a decrease in the strength of the shore deposits.

Beaches: energy absorbers

A beach is the best natural form of shore protection. The wider the beach, the better. When waves reach the beach and begin their run-up, the beach slope and sediment work together to dissipate wave energy, and thus the wave itself (fig. 2.10). This idea is well founded, and can be demonstrated easily along the Lake Erie shore. Wherever wide beaches (beaches 50 feet wide or more) front the shore, erosion rates are very low or nonexistent because waves do not reach the backshore to create unstable slope conditions. But during periods of record-high lake levels, as in the early 1970s, even fairly wide beaches became submerged or narrowed. As a result, waves once again reached the backshore slopes and caused accelerated erosion.

The link between lake level, waves, and beaches with respect

Wide beach

Narrow beach

2.10 Effect of beach width on shore erosion. (A) Wide beach protects the shore by absorbing wave energy. (B) Narrow beach allows waves to attack shore.

to shore erosion should be clear at this point. A small wave reaching the land at the back of the beach during a high lake level can erode more than a large wave at a low lake level because the energy of the large wave will be expended on the beach, not the land at the back of the beach. The key factor is whether the wave reaches across the beach to erode the toe of the slope.

Common questions about beaches

Where does the beach sand come from?

Lake Erie's beach sand is derived largely from shore erosion. Erosion of the glacial till (the "boulder clay") and rocky shores provides the sand grains of quartz, feldspar, and the rock fragments that make up most of a typical Lake Erie beach. As the slope or bluff collapses, wave erosion breaks down the debris that has accumulated at the base of the slope. The finest particles, the clay and silt, are transported by waves and currents into the deeper waters of the lake. These fine particles give the color to the brownish plumes in the water that are evident in the nearshore-offshore zones during storms. The more massive sand and gravel-size particles remain, at least temporarily, on the beach.

Little sand is derived from the streams flowing into Lake Erie because these streams generally are too weak to transport sand and gravel. Their lower reaches (mouths) were drowned by the post-Ice Age rise in lake level. The embayments at the stream mouths cause the stream currents to lose their strength before they reach the lake (e.g., the Maumee, the Cuyahoga). The result is deposition of sand and gravel on the stream bed upstream from the mouth. Whatever sand and gravel does reach the lake is usually unceremoniously dredged from the harbor channels and deposited offshore or in diked disposal areas.

Once the sand is separated from the silt and clay by the waves it is subject to a complex set of processes that tend to sort and transport it. Sorting refers to the separation of grains based on differences not only in size, but in shape and weight. The result of the sorting process can sometimes be seen as concentrations of brownish-red and black sand along the landward margins of some beaches. These sands are commonly called heavy mineral sands because they are made up mainly of high-density minerals such as garnet and ilmenite. These somewhat finer heavy minerals behave in the same way as the larger, lighter grains of quartz and feldspar. Both heavy and light minerals are carried by the strong wave uprush, but sorting during the weaker wave backwash transports mostly the lighter grains, leaving a concentrate of the finer, heavier grains on the backbeach.

Why do beaches change?

Graphic examples of the dynamic nature of a beach can be seen whenever a change in wave conditions or lake level occurs. One day the beach may have a smooth, uniform slope. The next day it may be characterized by a series of ridges. Generally though, the biggest changes in beaches take place during lake storms. Beaches are generally narrower and steeper during and after storms because the finer sand is transported offshore. This effect is not totally negative, however, with respect to shore erosion. As the sand is transported offshore, it builds up the nearshore bottom. This shallower bottom slows the storm waves due to friction between the wave and the bottom, and the waves lose some of their energy before they reach the shore. Similarly, sandbars sometimes build up offshore during storms. These bars trip incoming waves causing them to break before they reach the beach.

Sand carried offshore during times of storms is frequently transported back onshore by fair-weather waves. The sandbars formed in the storm often later "march" shoreward and the inner bar "welds" onto the beach. This produces a narrow troughlike feature that may form a shallow elongated pool at the edge of the beach.

How does beach sand move?

The largest volumes of sand moved are not in an onshore-offshore direction. Longshore or shore-parallel sand transport is more important in many areas. It is this process of longshore transport of sand (or more correctly the lack of it) that has contributed to many of the problems confronted by the lakeshore dweller. Longshore transport of sand is a surf-zone phenomenon, i.e., it occurs landward of the outermost breaking waves. If a wave breaks at an angle to the beach (the wave crests are usually not parallel with the shoreline) a portion of the breaking wave's energy is directed along the beach. Sand grains stirred up by the breaking waves flow like a stream in the direction of the longshore current (fig. 2.8). Moreover, sand grains are carried up the beach at an angle, then carried straight down the beach slope by the swash, and caught by the next incoming wave and again carried up the beach at an angle. The transport path of an individual grain of sand on the beach face is a sawtooth pattern in the direction of beach drift. Along rockbound stretches of the Lake Erie shore there is little or no sand, and these processes are correspondingly less important.

What are the directions of beach sand transport?

Because of variable wind and corresponding variable wave directions, the transport direction changes from time to time, causing reversals in sand transport. In general, however, a net transport direction results. This means that over a period of time there will be transport of sand in a prevailing direction along a given stretch of shore. This net transport direction is sometimes related to the orientation of the larger storm waves because sand transport is greater during storms than during fair weather. As noted earlier, the exposure of the shoreline is an important factor in determining wave direction and size and, therefore, the direction of longshore sand transport.

Along the Lake Erie shore, most of the sand transport is

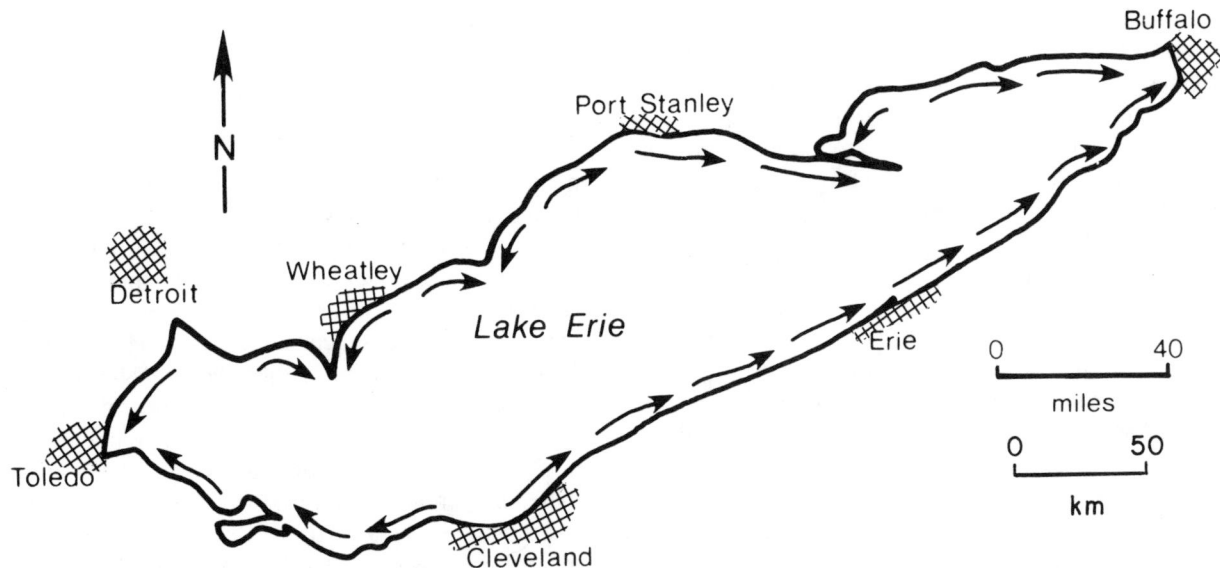

2.11   Net longshore sand transport directions for Lake Erie.

from west to east because of the prevailing westerly winds and southwest-northeast orientation of much of the shore (fig. 2.11). The orientation of certain shoreline features can be used to demonstrate the directions of sand transport. Presque Isle and Long Point spits (fig. 1.1) grew in easterly directions. The sizable accumulations of sand on the west sides of jetties (and the corresponding lack of sand on the east sides) from Port Glasgow east along the Canadian shore, and from Cleveland east along the U.S. shore all point to net west-to-east transport. The jetties have acted like dams on the beaches, holding back the flow of sand like a river's dam holds back water. West of Port Glasgow and Cleveland the longshore transport is largely reversed because the winds blowing from the east generate larger and more powerful waves as a result of greater fetch, even though the more frequent winds are from the west. This difference results in less frequent, but greater longshore sediment transport to the west.

The Michigan shore is unique in that its north-northeast trend and wave climate combine to generate a net transport to the south from Detroit to Toledo (fig. 2.11). The orientation of the sand spits at the Raisin River and the Woodtick Peninsula result from this north-to-south sand transport.

What do beaches tell us?

Because beaches are so important in terms of shore protection, longshore transport of sand is an important process. If the supply of sand in the longshore system decreases, the beaches become narrower because there is an insufficient incoming sand supply to replace the sand moving away.

Changes in beach width occur even if the balance of longshore sand is constant. As previously mentioned, storms commonly erode the beaches and transport sand both along the shore and offshore. More normal fair-weather wave processes commonly lead to the deposition of sand on the beaches. The most persistent changes, however, are due to the seasonal and annual lake level fluctuations. The one-foot drop in lake level between late spring and early winter generally increases the width of the beach by five to ten feet. The longer-term fluctuations, with a drop in lake level of three to four feet, cause increases on the order of tens of feet. The increase is largely related to beach slope. A beach with a gentle slope will show a much greater change in beach width than a beach with a steep slope for a given fluctuation in lake level.

A natural balance: the dynamic equilibrium

Beaches can be thought of as the end product of a set of natural laws, a dynamic system where a kind of balance is maintained among the factors that determine the beach's character. The main factors are: waves, lake level, sand supply, and the beach profile. The relationship among these factors is a natural balance, a dynamic equilibrium; when one factor changes, the others adjust accordingly to maintain the balance. Thus during fair weather or lower lake level, the beach widens and in some areas additional sand may be blown into dunes. When the lake rises or storm waves strike, the system retreats and the sand is redistributed in a profile to conform with the new set of energy conditions. There may be a lag time before the system comes into equilibrium, especially after a rapid rise in lake level. During such a time, humans often enter the system and construct shore protection (static) structures. When humans enter the system incorrectly, as we often do, the dynamic equilibrium continues to function and the results may be harmful to our best interests. In the process of protecting one house or property, we may cause damage to or the destruction of another. The price of saving buildings by constructing seawalls or similar structures may be the loss of the recreational beach. Nature's dynamic equilibrium and static structures are often a poor mix.

3 Man and the shoreline

Coastal engineering has had a tremendous effect on the shore of Lake Erie. This effect can be seen clearly by comparing lake charts and topographic maps from the 1800s, when man was just beginning to develop the shore, to present-day maps and charts of the developed shoreline. The harbor jetties, which were built to maintain the channels at the major stream mouths, caused the first significant man-made shoreline changes. Jetty construction was followed in the 1900s by the installation of shore stabilization structures to protect private property from erosion and flooding. Like the jetties, these structures have had both negative and positive effects. In many cases the structures have failed to deliver the expected property protection.

Experience with harbor structures and various shoreline stabilization approaches is now sufficient to establish more realistic expectations as to the results and costs of such structures. No property owner, developer, planner, or community official should opt for shoreline stabilization until he or she has considered questions such as: What effects will the proposed structures have on the shore, both in general, and for specific types of structures? Will the benefits really outweigh the harmful effects? Is the structure or project worth the true cost (initially as well as over the longer term)? What are the environmental and aesthetic trade-offs?

Harbor structures

Jetties are built to keep navigation channels open. The structures channel stream outflow, and reduce the rate at which the channel fills with sediment that normally would come from nearby beaches via longshore transport. Many jetties were built in the early 1800s and their effects on shoreline shape and beach width have been significant.

The Fairport Harbor structures at the mouth of the Grand River offer a good example of the effects of harbor structures. The first Grand River jetty built in 1827 was about 300 feet long. By 1831 the west jetty had been lengthened to 1,300 feet and the east jetty to 1,400 feet. In the early 1900s breakwaters were constructed with the west breakwater extending about 4,000 feet out into the lake, subparallel to the west jetty. The east breakwater extended about 6,800 feet parallel to the shore and perpendicular to the jetties (fig. 3.1).

Along this stretch of Lake Erie shoreline, overall sand transport is from west to east. The jetty-breakwater system interrupts the movement of sand along the shore. As a consequence, sand has accumulated on the jetties' west side, whereas the east side has undergone sand starvation. Such starvation results in the narrowing or loss of the downdrift beach. The sand moves along the west side of the jetty in a lakeward direction into deeper water, as well as into the harbor channel, and thus is lost from the beach zone.

The shoreline west of the Grand River structures has changed

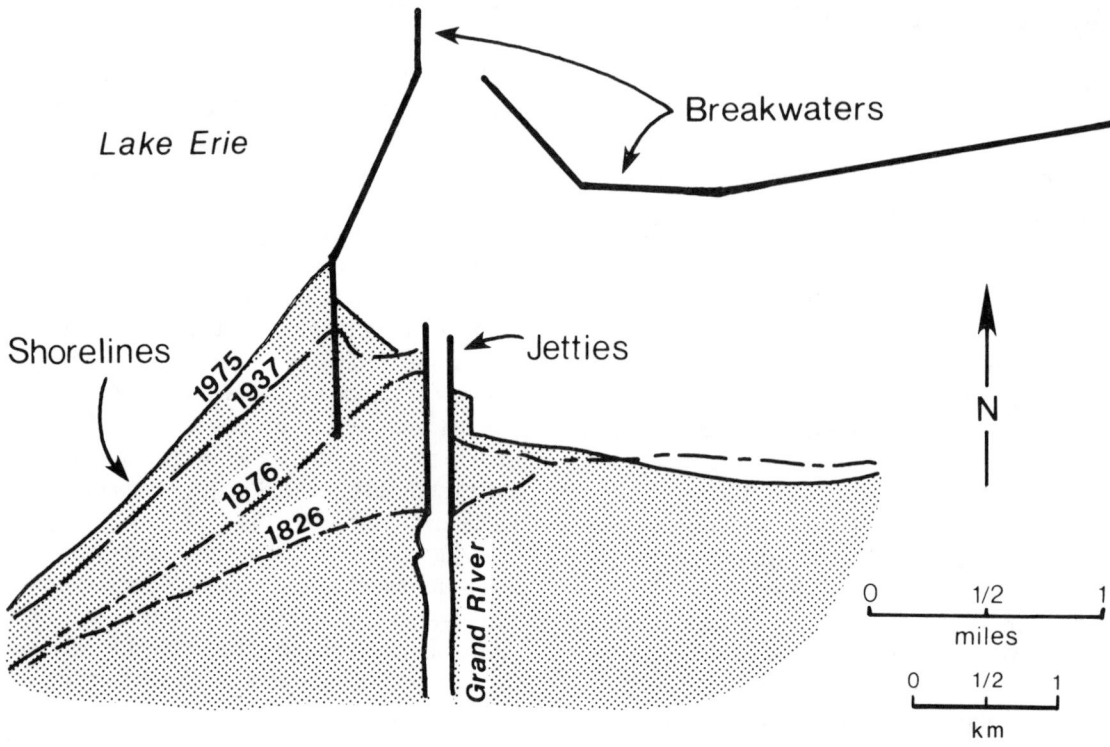

3.1 Jetties, breakwaters, and historic shoreline changes at Grand River, Ohio (Fairport Harbor). Modified from fig. 12 from *Coastal Geomorphology and Geology of the Ohio Shore of Lake Erie*, by C. H. Carter, D. E. Guy, Jr., and J. A. Fuller (reference 21, appendix B).

from a nearly east-west orientation in 1825 to a northeast-southwest orientation in 1975; but the orientation of the shoreline east of the structures has stayed about the same (fig. 3.1). A comparison of beach widths in 1876 to beach widths in 1968, times of about the same water level, shows that beaches have grown markedly wider for about a mile to the west of the structures. In contrast, beach widths have narrowed for about two miles to the east of the structures.

The most profound change is adjacent to the west jetty-breakwater, where the shoreline has advanced lakeward about 2,000 feet since 1827. Comparison of erosion rates between an early (1876–1937) period and a late (1937–1973) period shows a decrease in erosion rates for about a mile west of the structures. Little change has occurred in the erosion rates for the first mile and a half east of the structures (the east breakwater and an extensive system of seawalls and groins have helped protect this stretch of shore), but a marked increase in the erosion rates (two to four times the long-term rate) is typical for the next two miles to the east (fig. 3.2).

In general, the increased beach width to the west of the structures has led to a decrease in wave energy reaching the shore; the wide beach acts as a buffer between the waves and the backshore. However, the decreased beach width to the east of the structures has led to an increase in wave energy reaching the shore. The differences in both beach widths and erosion rates can be related directly to the Fairport Harbor structures. By trapping sand on their updrift (west) side from the longshore

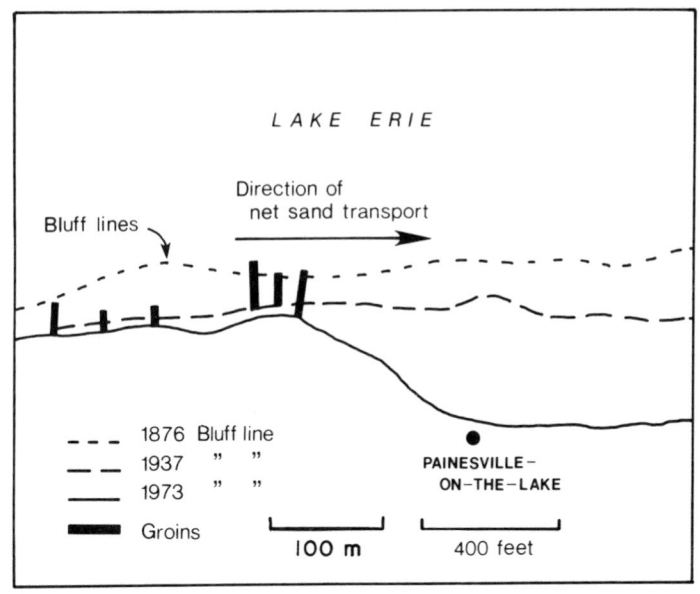

3.2 Shore erosion east of Fairport Harbor. The fairly uniform distance between the 1876 and 1937 bluff lines indicates parallel retreat of the shore at an average rate of about 2 feet per year. The nonuniform distance between the 1937 and 1973 lines indicates nonparallel retreat of the shore at an average rate of less than 1 foot per year along the western 800 feet of shore, and 10 feet per year along the eastern 800 feet of shore. Groins, which trap sand, helped stabilize the western section, but caused accelerated erosion along the eastern section.

sand transport system, the structures have helped protect the shore to the west by building wider beaches; and by blocking the transport of sand to their downdrift (east) side the structures have caused the beaches to become narrower. In turn, shore erosion has accelerated. The overall effects on both sides of the structures are not equitable. Erosion has been accelerated along downdrift stretches that are two to four times longer than the updrift stretches where erosion has been decelerated!

The Fairport Harbor jetty-breakwater system offers a dramatic example of shoreline change due to harbor structures along the Lake Erie shore. This example, however, is not an isolated case. The effects of similar structures at several other Lake Erie harbors, although not as great in magnitude as the Grand River, are basically the same (e.g., Port Stanley and Port Burwell along the Canadian shore, and Huron, Vermilion, Ashtabula, and Conneaut along the U.S. shore). Almost invariably one side is protected by the build-up of beach sand, and the other longer side is left unprotected by the loss of beach sand.

Shore protection structures

Shore protection structures are built to protect shorefront property threatened by erosion. Such structures are designed to hold the shoreline in place by dissipating or reflecting wave energy, by armoring the shore against waves, or by trapping sand. Although many variations exist both in design, and in construction material, there are three principal types of shore protection structures

used along the Lake Erie shore: breakwaters, seawalls, and groins. Like jetties, these structures usually affect the adjacent beaches. The number of such structures continues to grow at a phenomenal rate. For example, in Ohio alone in the late 1930s there were about 1,800 structures, and by 1973 there were about 3,600 structures. Where such structures are numerous and nearly continuous, evidence suggests that they are having a major effect on the shore (fig. 3.3).

Breakwaters

Breakwaters protect the shore by intercepting waves and reducing their energy before they reach the shore. Reduced wave energy allows sand to be trapped in the "shadow" of the breakwater, building a protective beach (fig. 3.4). The obvious benefit is reduced erosion and beach growth behind the breakwater. This protection, however, comes at the neighbor's expense. By interrupting the wave pattern, the breakwater reduces the strength of the longshore current allowing deposition of sand behind the structure. Moreover, this sand now acts as a barrier to additional sand transported by longshore currents and leads to narrower beaches and greater erosion along the downdrift shore (fig. 3.4). This effect can be reduced by artificially bypassing sand to the downdrift shore, but such systems add to the breakwater's design and operation costs. In addition, the breakwater itself may cause scour on the lake side of the structure, leading to its failure. Sometimes wave modification by the breakwater may intensify wave energy on nearby beaches adding to the erosion problem.

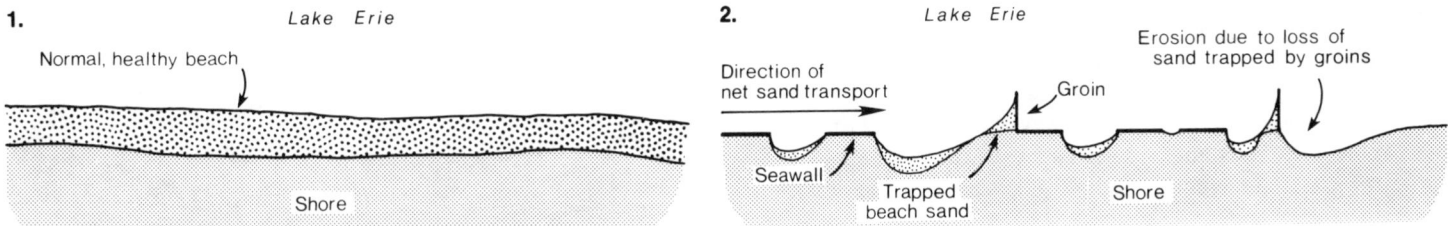

3.3 Cumulative effect of seawalls and groins. The structures contribute to the loss of the beach and shoreline aesthetics.

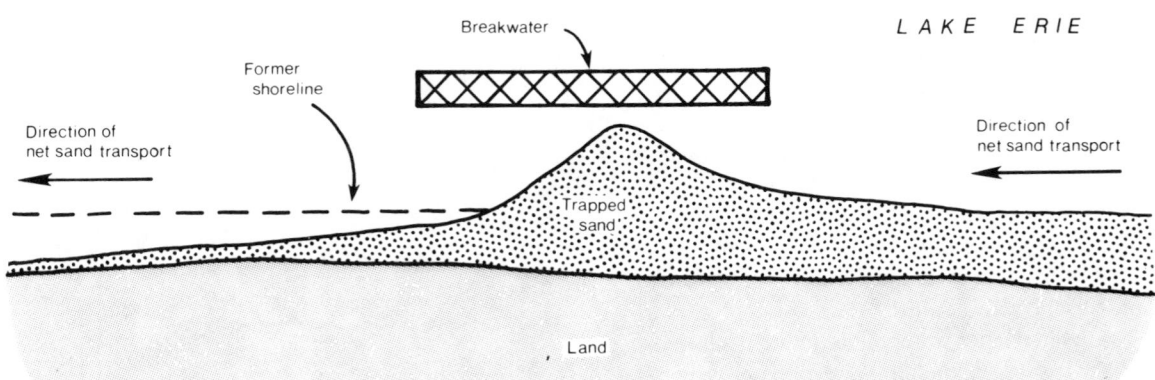

3.4 Map view showing the effects of a breakwater. Note entrapment of sand behind the breakwater and loss of sand to the downdrift beach.

Seawalls

Seawalls armor the shore behind them (fig. 3.5). These structures are designed to protect the property behind them by reflecting or absorbing the energy of breaking waves, i.e., making the shore more wave resistant. Seawalls are constructed of a wide variety of material including riprap (broken rock or boulders), concrete, steel, wood, and sandbags. They vary in size from massive walls to protect large developments (fig. 3.6) to small individual property structures (fig. 3.7).

The general category of seawall structures includes *revetments*, which armor the shore to reduce wave erosion, and *bulkheads*, which are walls at the back of the beach intended to hold the land in place. Like seawalls, these structures are built of a variety of construction materials, and their design usually calls for modification of the landward slope if they are to have a chance of protecting the property behind the structure.

Seawalls can successfully protect property (while the walls last), but they also can cause problems. The smooth, vertical walls of the structures deflect wave energy downward in front of the wall or laterally along the wall. As a result, sediment is eroded from in front of the wall at its toe. Removal of this support may contribute to the wall's failure. The erosion of sand in front of the wall results in deeper water. This, in turn, allows bigger waves to strike the wall, expending greater energy, contributing to greater erosion at the toe of the structure until the wall fails (fig. 3.5). This is why small walls tend to be replaced

1.

Normal, healthy shore

2.

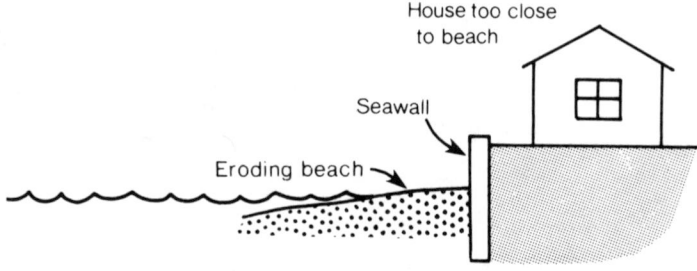

House too close to beach

Seawall

Eroding beach

3.5   The life of a seawall. Wave energy reflected off the wall removes sand from in front of the wall, leading to scour at its base, water

3.

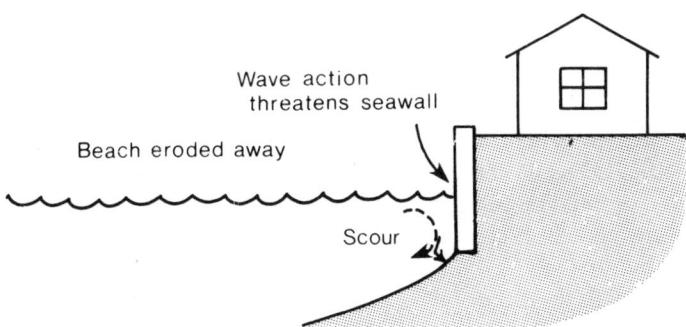

Wave action
threatens seawall

Beach eroded away

Scour

4.

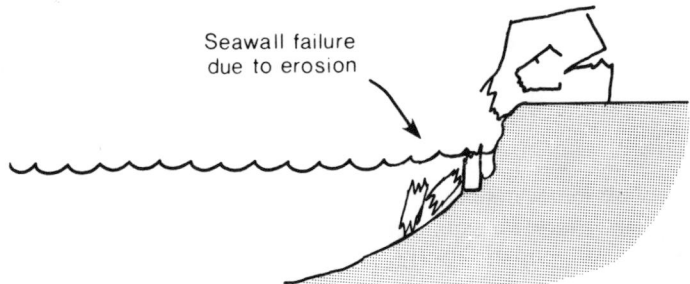

Seawall failure
due to erosion

deepening, and accelerated longshore transport. These effects lead to structural failure.

3.6 Large seawall at Euclid, Ohio. Note the lack of a beach in front of the seawall.

with larger walls, and why one sees double wall structures along some Great Lakes shorelines.

In the Great Lakes, many seawalls are placed at the front of the beach or at the water's edge, and are intended to trap sand as well as to dissipate wave energy. Breaking waves transport sand as well as water over or through the wall (some designs leave slits in the walls), and some of the sand remains behind the wall to build a *perched beach*. Sand also may be placed behind the wall

3.7 Typical small seawalls on the Lake Erie shore.

at the time of construction. If sand is not trapped, or if the back-fill is lost, the wall is likely to fail. Walls may fail to prevent erosion when water levels exceed effective design heights during storms. Large storm waves are capable of lifting or moving rip-rap, and such waves exert great force against any structure, enough in some cases to cause damage or destruction.

Accelerated erosion usually takes place at the unprotected ends of the wall. This effect creates a reason to extend existing walls or build more walls. In addition, deflected wave energy speeds up longshore transport, removing the dwindling amounts of sand.

Seawalls also contribute to erosion by cutting off the supply of beach sediment normally derived from shoreline erosion. Erod-ing bluffs are the principal source of beach sediment for most of the Lake Erie shoreline, so for the most part seawalls force a trade-off: shoreline retreat is retarded and buildings are saved, but beach width and continuity are reduced, sometimes drastically.

Dikes

Dikes are somewhat like seawalls and are found along extensive lengths of shore; for example, the dike that fronts the Ottawa National Wildlife Refuge (Little Cedar Point) at the west end of the lake is nearly five miles long. The dikes protect the easily erodible marshes from waves that would quickly convert marsh-land to part of the lake. The natural sand barriers that fronted the marshes were disappearing because of the effects of man-made structures. Long before the construction of the dikes, large quantities of sand had been moved intermittently by longshore transport during storms until deposited at natural barriers such as Maumee Bay and Point Pelee. Diminished sand supply and the effects of seawalls and groins on adjacent beaches meant that the protective sand barriers could not rebuild after high lake level periods. The dikes were built because the less-than-normal sand supply resulted in an increasingly greater erosion toll on the barriers with each successive high water.

The dikes do preserve the natural shoreline, but in an unnatural way. This presents preservers of the natural system with a dilemma. The Division of Natural Areas and Preserves of the Ohio Department of Natural Resources is confronted with such a problem at Sheldon's Marsh State Nature Preserve just west of Huron, Ohio. The preserve includes part of the easterly end of Cedar Point spit, a bay mouth spit that has built westward from Huron to front the Sandusky Bay mouth. The spit in the preserve was breached in the early 1970s, and since then the unprotected spit has continued to be pushed back into the marsh as sand is washed over the barrier during storms. Presumably the barrier eventually will be completely eroded leaving an embayment bordered by built-up stretches on either side of the preserve. The preserve managers would like to maintain the integrity of the barrier as well as the marsh, but are reluctant to do anything that would be considered unnatural.

One can argue that unnatural causes brought about the demise

of the spit—after all, the Huron jetties have had a profound effect on the entire stretch of shore west of Huron. The jetties, by trapping sand from the east-to-west longshore transport system, have caused accelerated erosion from the Huron River to the tip of Cedar Point. Moreover, accelerated erosion along the west side of Huron (because of the jetties) led to the construction of shore protection structures. These shore protection structures have reduced the amount of sand moving along the shore, because they prevent erosion of the shore, which in turn provides sand for the beaches and the longshore transport system. These conditions have created a wholly unnatural setting for the preserve.

On the other hand, the build-up of sand on the west side of the Fairport Harbor structures has created the setting for Headlands Dunes State Nature Preserve. Here sand dunes and related flora rare to Ohio developed because of the entrapment of the sand by the harbor structures. Although the setting is unnatural in the sense that it developed because of the jetties, one can argue that the setting would likely have been preserved somewhere along the shore even if the white man had never come.

Groins

Groins are wall-like structures built perpendicular to the shore (fig. 3.8). These structures are designed to trap sand and slow the rate that it is transported along the shore. In cases of low sand supply, sand may be emplaced artificially between groins at the time of their construction. This procedure, however, does

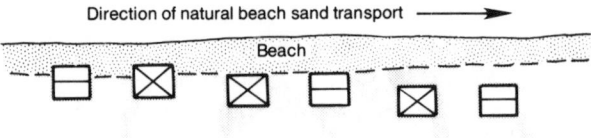

Direction of natural beach sand transport ⟶

Beach

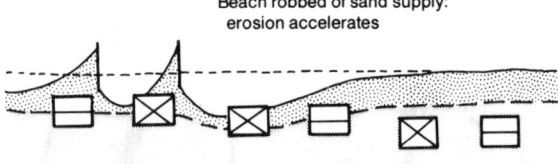

Beach robbed of sand supply:
erosion accelerates

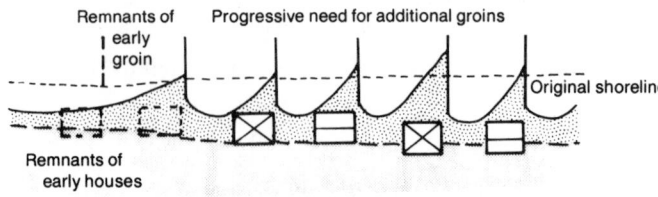

Remnants of early groin   Progressive need for additional groins

Original shoreline

Remnants of
early houses

3.8 Evolution of a groin field. Groins built to protect threatened houses rob downdrift beaches of sand, leading to accelerated erosion and the installation of more groins. The problem is simply shifted along the shore.

not eliminate the effect of the groins in modifying the longshore transport of sand.

In most cases, groins act as small jetties in that the updrift shore will be partly protected by trapped sand. However, by interrupting the supply of sand moving along the shore, the groins starve the downdrift beaches, causing them to become narrower or to disappear. Accelerated erosion then occurs along the downdrift stretches on a length two to four times longer than the updrift stretches that experience decelerated erosion because of wider beaches.

Like seawalls, groins may contribute to their own demise. In addition to fostering downdrift sand starvation, the structures may cause sand to be diverted into deeper offshore waters, or may concentrate wave energy, which speeds up erosion. Loss of effectiveness often occurs when the land at the shoreward end of the groin is eroded from the groin (*detachment*) allowing sand to bypass the groins. The structures are often damaged by ice, slope forces, wave impact, or undercutting as a result of wave scour.

Beach nourishment (replenishment)

Beach nourishment means adding sand to a beach. This non-structural or "soft" stabilization approach is used along the Pacific, Gulf, Atlantic, and Great Lakes coasts of the United States. The intent is to increase the width and length of the beach for shore protection and recreation. The concept is a reasonable

A

B

3.9 Beach nourishment project at Presque Isle State Park, Erie, Pennsylvania. (A) Beach nourishment in progress. (B) Beach after nourishment is complete. Photographs courtesy of the Buffalo District, U.S. Army Corps of Engineers.

one, but one that has met with mixed success (reference 76, appendix B).

The biggest drawback to beach nourishment is cost. For example, the nourishing of ten miles of beach along Miami Beach, Florida, cost $65 million. Overall, the cost is usually closer to $1 to 2 million per mile on U.S. ocean beaches. Moreover, the replenished sand along most open-ocean beaches disappears rapidly, and it is necessary to repeat the nourishing from time to time ad infinitum. Replenished open-ocean beaches tend to erode more rapidly than the natural beaches they replace.

On Lake Erie four beach nourishment projects have been undertaken by the U.S. Army Corps of Engineers: three small projects—Lakeshore Park, Ashtabula, Ohio; Lakeview Park, Lorain, Ohio; and East Harbor State Park near Marblehead, Ohio—and a larger project at Presque Isle State Park, Erie, Pennsylvania (fig. 3.9). The small beach nourishment projects appear to be losing sand slowly because they are in relatively sheltered areas that are not subject to large waves and strong longshore currents. The project at Lakeview Park is particularly interesting because the sand used in the project was dredged from the Pelee-Lorain moraine about six miles offshore of Lorain.

The Presque Isle project is the largest beach nourishment project in Lake Erie. About 7.1 million cubic yards of sand were spread on the beach at Presque Isle over a 30-year period at a cost of about $16.8 million. Here, because of the strong longshore currents generated by wind-driven waves from the west, the sand was rapidly eroded and transported along the beach to the east or offshore. Because of the project cost, other measures such as offshore breakwaters are being considered to save Presque Isle and its enclosed harbor from erosion.

Another form of beach nourishment known as sand bypassing may be a reasonable approach for stretches undergoing accelerated erosion because of harbor jetties (e.g., Fairport Harbor, Ohio). In this method, sand is dredged from the updrift side of the structure and deposited on the downdrift side. In essence, the transport of sand along the shore is maintained and beaches are reestablished along the downdrift shore from the harbor structures, providing shore protection and recreation. Unfortunately, most of the sand dredged from Lake Erie's harbors is presently dumped offshore or in diked disposal areas, and as a result is lost from the coastal system (i.e., the supply of sand along the shore is reduced).

Although closer to the natural system than seawalls, groins, or breakwaters, beach nourishment is not a permanent panacea to the erosion problem. If replacement sand is finer than the original beach sand, it will erode more quickly than the original sand.

If the nourishment does not conform to the natural beach slope, greater wave energy may attack the replacement beach. In any event, the erosion that led to the need for nourishment will require continuing additional placement of maintenance sand. As the Army Corps of Engineers has noted, sand is usually a diminishing resource.

Structures on the shore: a growing problem

As noted in chapter 2, the shape of the shoreline is controlled by the lake level, the wave regime, the sediment or rock type, and the sand supply. An equilibrium or balance exists among these factors. When any one of them is altered, the shape (profile) of the shoreline will change as it moves toward reestablishment of a new balance or equilibrium. Stabilization structures, and even beach nourishment, have a significant impact on this equilibrium. The history of human interaction with the Lake Erie shore shows that adopting a structural stabilization approach (by building shore protection structures) has the short-term gain of temporary protection of property, traded off against intermediate and long-term losses including maintenance and replacement costs, loss of beach, and aesthetic loss. Table 3.1 summarizes the advantages and disadvantages of methods used to reduce erosion.

Studies by the Ohio Geological Survey indicate that the overall effect of shore protection structures in Ohio has been to reduce the rate of shoreline retreat (reference 73, appendix B).

Table 3.1 Methods used to reduce erosion and erosion loss

Move Building	Beach Nourishment	Landscape Shore	Seawalls[1]	Groins	Breakwaters
ADVANTAGES					
Reduces threat of loss	Reduces erosion by increasing beach width	Increases stability of slope	Reduce erosion by armoring shore	Reduce erosion by trapping sand and increasing beach width updrift of structures	Reduce erosion by reducing wave energy reaching shore
Allows natural shore processes to operate	Allows natural shore processes to operate	Allows natural shore processes to operate			Increase beach width behind structures
May be less expensive than shore structures	May increase attractiveness of shore	May increase attractiveness of shore	Increase stability of slope by adding resistance to toe		
	Increases beach width downdrift				
DISADVANTAGES					
Does not reduce erosion	Periodic replenishment usually necessary	Does not materially reduce erosion	Wave reflection erodes beach in front of seawall	Impede the longshore transport of sand causing erosion downdrift	Impede the longshore transport of sand causing erosion downdrift
Sufficient area must be available for move	Expensive	Sufficient area must be available for landscaping	May limit access and recreational use of beach	May limit beach use	Need periodic maintenance
Building must be movable	May impact on ecosystem	Expense variable	Need periodic maintenance	Need periodic maintenance	More expensive than seawalls or groins
May reduce lake-view or vista			May reduce attractiveness of shore	Need abundant sand in longshore system	May reduce attractiveness of shore
			May reduce sand supply to beach through shore erosion	May reduce attractiveness of shore	

1. Includes bulkheads and revetments

Along some segments of the shore, particularly downdrift of groins, there is accelerated erosion, but the overall, net effect of the nearly 4,000 shore protection structures is decreased erosion. The number of such structures along the Ohio shore has increased from about 60 in the mid-1880s to 1,800 in the late 1930s, to about 3,600 in 1973. This pattern is typical for developed shorelines along the rest of Lake Erie and other Great Lakes. Clearly structural shoreline stabilization is viewed as the solution to the erosion problem. But what is the true cost of this "solution"?

The most obvious cumulative effect of this tremendous increase in structures has been to modify the distribution of the beaches and to change the shape of the shore. In the 1800s the shore of Lake Erie had a nearly continuous beach, which was even used as a highway. By the mid-1900s this beach, particularly along the most built-up stretches, was narrowed, broken, and fragmented as the jetties and groins modified and interrupted the longshore transport of sand. The numerous seawalls, by reflecting and deflecting wave energy, caused accelerated erosion of the beaches in front of them, and at their ends, adding to the overall beach loss.

The shape of the shore changed from relatively smooth and gently curving to irregular and uneven (fig. 3.3). Differences in erosion between the segments protected by the man-made structures and the unprotected segments led to this more irregular shore, one characterized by artificial armored headlands and retreating embayments without protective structures. As a result, wave energy reaching the shore is more irregularly distributed, reducing longshore transport rates and further interfering with sand transport. The effect is additional beach loss, local accelerated erosion, and the perceived need for more structures.

Along irregular shorelines most of the wave energy is expended on projections of the shore that extend lakeward, in this case the armored headlands. If insufficient energy reaches the embayments, then the beaches in the embayments will undergo little if any sand transport. The sand essentially will be trapped in the embayments. As sand builds up in the embayments, increasing beach width, slower erosion rates will result because less wave energy will reach the shore. If this process continues, some stretches of shore will consist of headlands protected by massive arrays of shore protection structures (mostly seawalls or breakwaters as there will be little or no sand in the longshore transport system for groins to trap) and embayments fronted by pocket beaches (fig. 3.10).

3.10 Armored headlands separated by embayments with pocket beaches.

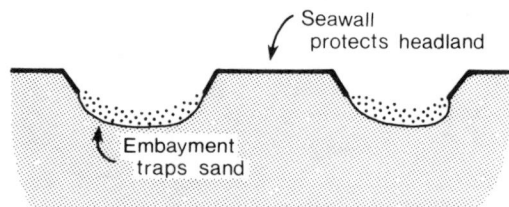

Coastal development and coastal engineering: five axioms

Our experience and the experience of countless others suggest at least five general axioms that apply practically everywhere that development takes place along the shoreline.

1. *Erosion is not a problem until the shore is developed.* Shore erosion is a natural, ongoing process. There may be periods when little or no erosion takes place along the shore, but in the long run, erosion inevitably takes place. Until development, erosion is not perceived as a problem. A naturally eroding, undeveloped shoreline poses no problem or threat. Problems arise after the shore is developed, and people attempt to combat natural processes and hold the shoreline in a fixed position. The perception of lakeshore flooding as a problem also comes only after shoreline development.

2. *Coastal construction causes coastal change.* The shoreline is a dynamic setting where change is the rule rather than the exception. Coastal construction must be regarded as a precipitator of changes that invariably have a marked effect on the natural setting and processes. Jetties and groins alter the longshore transport of sand and thus affect the width and lateral continuity of the beaches. Seawalls, bulkheads, revetments, and breakwaters reflect and deflect wave energy, and in so doing alter the transportation and deposition of beach sand, and reduce the supply of sand coming into the beach zone.

3. *Stabilization structures protect property, not beaches.*

Shoreline stabilization structures are sometimes sold on the premise that it is the beach that will be protected. In fact, shoreline projects are carried out to protect shorefront property, not beaches. Structures such as groins and seawalls often result in the destruction of the beach and accelerated erosion of adjacent unprotected shoreline.

Large-scale shore stabilization projects are in the interest of a relatively small number of shorefront property owners rather than the public. Yet beach property owners apply pressure for the spending of public funds to protect privately owned shore. Do the personal interests of shorefront property owners warrant large expenditures of public money for shoreline stabilization, considering that the beach (public domain in some areas) will be lost in the process?

Even small structures placed at the expense of property owners may impact negatively on the shore. Beach loss, accelerated erosion of neighboring shore, impediments in the water line right-of-way, and aesthetic loss are costs to the public.

The grand example of where this leads is well illustrated by the long-stabilized New Jersey shore. Miles of "well-protected" shoreline are without beaches. For communities such as Sea Bright, Long Branch, Deal, and Monmouth Beach, the "beach" is usually a pile of rubble from destroyed seawalls and groins (reference 115, appendix B).

4. *In the long run, the cost of shore protection structures is commonly greater than the value of the shore property to be protected.* Shore protection structures on Lake Erie cost from a

few tens of dollars per linear foot along the low, sheltered clay banks of Sandusky Bay to well over a thousand dollars per foot for more involved projects that include breakwaters and beach nourishment. For example, the U.S. Army Corps of Engineers project at the west end of Lake Erie including a sand beach with offshore breakwaters and a revetment is estimated to cost about $1,000 per linear foot of shore protected. In general, one can realistically expect to spend at least $200 to $300 per linear foot for reasonably well-built (but still temporary) shore protection along much of Lake Erie. The cost is higher if the bluff is higher, so the cost of comparable protection along the central Canadian shore with its greater relief is much more than for the central U.S. shore. Naturally, structures need maintenance or replacement in time, which adds to the cost. During high lake levels and stormy seasons many of the smaller structures in the Great Lakes have been destroyed in less than one year. Therefore, costs are unpredictable. The U.S. Army Corps of Engineers, even with all their engineering expertise and experience, tried four different types of revetments to control lakeshore flooding along central Lake Erie at the mouth of the Chagrin River at a total cost of $1.5 million. The total value of the homes protected was only about $1.7 million.

Initial price estimates for shore protection often turn out to be unrealistically low for a variety of reasons. Maintenance, repair, and replacement costs are typically underestimated. Cyclic high lake levels or major storms are viewed not as inevitable, but as catastrophic acts of God or sudden strokes of bad luck for which one cannot plan and need not budget. The possibility of damage resulting from the shoreline structure itself is also ignored in most cost evaluations. Finally, costs are typically projected for one-time construction. Very few shoreline-stabilization projects would be funded at all if those controlling the purse strings realized that such "lines of defense" must be perpetual.

5. *Coastal shore protection construction, once started, is difficult to stop.* This truth is confirmed by the history of coastal engineering throughout the world. Once a shore protection structure is built, changes caused by the structure to the adjacent shore commonly result in the construction of yet another structure and so on. Moreover, the structures tend to steepen the offshore profile, reduce the sand supply, and accelerate erosion so that subsequent structures have to be bigger, better, and thus more expensive to remain effective. The engineering solution must be maintained indefinitely. The necessity of maintaining or rebuilding structures is quite apparent along the older Lake Erie communities. If you swim or take a small boat and poke around on the lake bottom adjacent to the shore you commonly will find debris from early engineering projects (e.g., boulders, concrete forms, steel rods, and beams) that failed or could not be maintained, which necessitated another set of structures.

History shows us that there are two situations that may terminate shoreline stabilization projects. First, a civilization may fail and no longer build and repair its structures. This was the case with the Romans, who built mighty seawalls. Second, a large storm may destroy a shoreline stabilization system so thoroughly that people decide to stop trying. Along the Great Lakes, however, such storm damage is usually met with a persistent resolve to hold the line, which results in ever-more-elaborate shoreline stabilization projects. The aforementioned debris of older structures is evidence of this response. Shoreline stabilization projects are viewed as the only alternative to failed shoreline stabilization projects.

Some implicit rules for formulating lasting solutions

The remaining chapters suggest some approaches for avoiding or reducing the conflicts between coastal development and natural hazards. However, a brief review of some rules is in order.

1. *Pick a safe site.* Safe is a relative term, but shoreline stability and the potential for coastal hazards can be evaluated. Chapter 4 provides a basis for safe site selection. In all cases, a prudent setback distance from the shore is in order. Michigan requires new houses to be set back 30 times the average annual erosion rate. The prudent homeowner will at least double that distance. Even slow erosion rates may bring houses to the brink of trouble, but long after the mortgages have been paid and the lives of the buildings lived!

2. *Design to live with the flexibility of the shoreline: be prepared to move back.* Be ready to move back rather than fight nature with a static line of defense. In many cases the cost of moving a house would be less than building, maintaining, or replacing protective structures over time.

3. *Consider buildings near the shoreline as temporary.* Although it may be difficult to accept or admit, there are times and conditions under which it is reasonable, economically or environmentally, to let nature take its course and to abandon a building to the lake.

4. *Base decisions affecting shore development on the welfare of the public rather than on the welfare of shorefront property owners.* This rule is not meant to be discriminatory or punitive. The nation faces a growing problem with development that ignores natural hazards. Policies should not encourage people to live in hazardous zones. Persons choosing to live in hazardous zones (e.g., along active earthquake faults, on the slopes of dormant volcanoes, on active floodplains, or along dynamic shorelines) for the associated benefits should expect to bear the risks of personal loss.

5. *Do not accept stabilization projects for shore protection as either a one-time or a permanent solution.* Shoreline stabilization, particularly shore protection structures, should be viewed as a last resort. Such stabilization is economically sound on a long-term basis probably only for metropolitan areas.

Questions to ask if shoreline protection projects are proposed

When a property owner or community is considering some form of shoreline protection, there is often an atmosphere of crisis. Buildings and commercial interests are threatened, time is short, an expert is brought in, and a solution is proposed. Under such circumstances, the right questions are sometimes not popular and are often not asked. The following is a list of questions to bring up if you find yourself a member of such a community, or if you are considering shoreline stabilization projects for your property.

1. Will the proposed project affect existing beaches? in 10 years? in 20 years? in 30 years?
2. How much will maintenance of the project cost in 10 years? in 20 years? in 30 years? in the longer term?
3. If the proposed project is carried out, what is likely to happen in the next big storm? in the next cycle of high lake levels?
4. What has been the erosion rate of the shoreline in the last 10 years? 20 years? since the late 1930s when the first coastal aerial photography was shot for Lake Erie? since the mid-1870s when the first accurate shoreline maps were surveyed by the old U.S. Lake Survey?
5. What will the proposed project do to the shorefront along the coast? Will the correction of one portion of the shore create problems for another portion?
6. If the proposed erosion project is carried out, how will it affect the type and density of future shorefront development? Will additional erosion projects be needed at the same time as the new development?
7. What will happen 10 or 20 years from now if the nearby stream or harbor entrance is dredged for navigation? if jetties are constructed or extended? if seawalls and groins are built on nearby property?
8. What is the 50- to 100-year environmental and economic prognosis for the proposed project if predictions of future lake level rise are accurate?
9. If stabilization (for instance a seawall) is permitted, will this open the door to seawalls elsewhere on the shore? (The answer to this historically has always been "Yes.")
10. What are the alternatives to the proposed project? Should the threatened buildings be allowed to fall in? Should they be moved? Should tax money be used to move them?
11. What are the long-range environmental and economic costs of the various alternatives from the standpoint of the local property owners? the local community? the entire shore? the citizens of the state or province and the rest of the country?
12. What will the alternatives be if future regulations are adopted that prohibit such structures (as has happened in North Carolina)?

If, after reviewing such questions, the decision is to proceed with a stabilization project, individuals and communities should follow existing guidelines for such projects.

4. Selecting a site on the Lake Erie shore

The high lake levels of the early 1950s and early 1970s with associated flooding and erosion problems returned to Lake Erie with record-setting levels in the mid-1980s. In fact, record high levels are expected to continue into 1987. The media is once again reporting on the plight of the shoreland owners who are caught between a rock and a hard place as accelerated shore erosion and record-breaking flood levels threaten, damage, or destroy lakefront homes and buildings. In spite of the significant problems and dangers inherent in living along the shore, people still want to live by the lake. Do the associated problems and dangers make coastal development a hopeless situation? We do not think so; but we feel strongly that coastal development must proceed in a careful, logical way.

Over the past 30 years, and particularly over the past 15 years, there has been a tremendous amount of research done on erosion and flooding along the Lake Erie shore (references 50–69, appendix B). The results document the changes and effects of erosion and flooding, and substantiate our intuitive knowledge of the major up-and-down fluctuations in lake level and associated flooding and erosion problems.

The facts speak for themselves. Marked intervals of high lake levels caused by above-normal precipitation recur on the order of every 10 to 20 years. At times of high lake level, lake flooding and erosion result in widespread damage and destruction to homes and buildings adjacent to the lake. In spite of this record of damage and destruction, renewed construction returns when the lake returns to a lower level. This construction often is in areas that sustained major damage just a few years before.

Because of widespread shore erosion and storm flooding few dwelling sites along the Lake Erie shore are without risk. Intelligent and careful planning, however, can result in the selection of sites with the lowest risk, thus making the coast a safer place to live. That's what this chapter is all about. If man chooses to confront nature at the shore, he should at least be aware of the possible consequences.

Nature's clues to danger at the shore

For the wary coastal dweller, nature holds many clues that can reveal much about the safety or vulnerability of a particular area near the shore. Although it helps to be an expert on coastal processes, expertise is not a requirement to carry out a prudent evaluation of the risk of coastal hazards. In fact, as one reads through the evaluation list below it is apparent that many of the indicators of site safety are simple common sense. On the other hand, some aspects of site choice can't be resolved by common sense alone. For example, one usually examines a site on a bright, sunny summer day. The lake is calm and waves are small or nonexistent. It is extremely difficult to imagine what the lake looks like during an intense fall storm or what the waves of that storm pounding the shore can do.

Another aspect of site choice where common sense often fails is the long-range view of man's impact. For example, a long seawall in front of large buildings may seem to coexist in perfect harmony with a broad beach, but 20 to 60 years from now the beach in front of the wall almost surely will be gone and the wall probably will have been replaced by a much larger structure.

Although the erosion rates cited in this chapter are well documented, some people are skeptical of them. Too often people fail to understand the significance of such erosion rates in terms of where the shoreline will be when their children or grandchildren inherit the house.

The wise landowner also knows that more than natural forces are at work. The politics of a community play a major role in determining how the community will interact with nature. Many American coastal communities with summertime populations in the tens of thousands are controlled politically by a few hundred year-round residents. Therefore, it's important to understand the politics of a beach community.

Apart from the political climate, the most important natural clues to site selection along the Lake Erie shore include relief, shore material, the natural shore environment, and existing shore protection structures.

Relief

Low, flat areas (areas of low relief) are flooded when water rises during a storm. These areas are also more subject to damage associated with the storm surge. Naturally, high areas are less prone to flooding and direct wave attack, so the prudent buyer will select a site above the lakeshore floodplain if possible. "High" relief along most of the Lake Erie shore is on the order of 10 to 15 feet above the long-term average lake level, but at the ends of the lake, and especially at the east end, where the surges are greatest, locating 10 to 25 feet above lake level is prudent. You can purchase a topographic map to find out the natural relief of the site, or less preferably, you can estimate by eye. However, relief alone is not a guarantee of low risk; the nature of the shore material is a crucial factor.

Shore material

All shore materials erode, but different materials erode at different rates. Rock (shale and carbonates) erodes most slowly, commonly at an average rate of less than one foot per year. Till (boulder clay) erodes at a rate of about two feet per year; laminated clay at about five feet per year; and sand at highly variable rates. (Sand is the least resistant shore material, but unlike other materials it can build out as well as erode back). These erosion rates are long-term averages and thus do not reflect the tremendous short-term variation related to low and high lake level intervals. Obviously, a buyer who has a choice should select the site made up of the shore material that erodes most slowly, but erosion is almost certain to take place, regardless of the site or material.

Natural shore environment

Two general environments characterize the Lake Erie shore. The barrier-marsh and clay bank environment makes up most of the shore at the west end of the lake, generally from Sandusky west to the Detroit River, along with Point Pelee, Rondeau, Long Point, and Presque Isle. The barrier-marsh consists of low (5 to 10 feet) sand ridges that parallel the lake and form a barrier between the lake and the marsh. This is the most unstable and dangerous environment along the lakeshore. The low relief allows storm surges to top the barriers, and the easily erodible nature of the sand that makes up the barrier allows for rapid destruction of lakeshore dwellings built on the barriers. The clay banks that are commonly associated with the barrier-marsh have low relief (5 to 10 feet) and are made up of thin layers of silt and clay that are easily weathered into small, angular blocks. The low relief allows storm surges to flood the banks, and the clay, once it is broken up by weathering, is easily eroded.

The rock or till bluff environment makes up the lion's share of the shore. Although the risks are not as clearly identifiable as along the barrier-marsh and lake clay shore (especially on a nice day during a low lake level interval), significant risks are often incurred along this type of shore simply because many people think that such a setting is invulnerable to erosion. A lakeshore site behind a wide (50 to 100 feet) beach backed by a stable, vegetated slope that has experienced little (if any) erosion in the last few tens of years is probably a relatively stable site. However, if the beach is narrow and backed by a bare slope, and if erosion has taken place in the last few tens of years, the property buyer should be cautious. The problem is particularly acute with the till shore because of the multifaceted problems associated with landslides. The problems are exacerbated on high bluffs or slopes because of the tremendous gravitational forces acting on the shore. However, the problem begins when waves cross the narrow (if existent) beach and attack the toe of the bluff or slope. This is just one reason why a beach is so important to coastal stability.

Shore protection structures

Nature creates a lot of headaches for people who live along the shore. But a choice to confront nature inevitably accentuates the natural hazards. Furthermore, we may not only increase the hazards from natural forces, we may create additional hazards.

The major people-created problems along the Lake Erie shore are related to man-made structures: harbor structures and shore protection structures. Both of these types of structures are discussed in chapter 3. Briefly, the harbor structures have had a major impact on the longshore transport of sand. This modification of sand transport has caused beaches downdrift of the structures to become narrower and the shore behind the beaches to erode at accelerated rates. Shore protection structures, some of which were built because of the harbor structures, have led to discontinuous, narrow beaches at great cost to individual home owners. Furthermore, these structures are usually quickly damaged or destroyed by the lake, requiring additional (and often continuous) expense for maintenance and new construction. Therefore, a property buyer evaluating lakefront property should regard the presence of shore protection structures as indicative of a potential problem.

The site: checklist for safety evaluation

The following is a guide to evaluating a potential site in terms of site safety. Find a shoreline area of interest later in this chapter, and using the general information given in chapters 1, 2, 3 plus the following checklist, evaluate the site.

1. Site elevation is above anticipated storm surge level.
2. Federal flood insurance is available for floodplain areas.
3. Building codes exist and are enforced.
4. Site is in an area of shoreline growth (accretion) or has a low erosion rate (less than one foot per year). Obviously, the farther the site from the edge of the bank or bluff the better.
5. In the case of rock, the bank or bluff is not markedly undercut (unless, of course, you don't mind losing the undercut shore to erosion in the near future).
6. In the case of till, the slope is fronted by a wide beach, covered with vegetation, and has a stable slope of about 30 degrees (unless you don't mind losing a few tens of feet of shore during a high lake level interval).
7. The year-round residents (who are the electorate) agree with your outlook on the future of the community.

Regional setting (reaches)

The first step in selecting or evaluating a site is to look at the regional character of the shoreline. Long distances (tens of miles) of shore (reaches) can be characterized in terms of their location, physical characteristics, and physical processes such as flooding and erosion. For example, the low, marshy western shore of the lake is more susceptible to flooding and erosion than the high, rockbound shore between Erie and Dunkirk that erodes slowly and never floods except at stream mouths. The reach descriptions that follow include (1) their physical nature and processes, (2) erosion or flood hazards, and (3) suggestions for hazard control or reduction.

Local setting (stretches)

Within a reach considerable variation in physical characteristics and processes occur. Shorter distances of shore (miles or frac-

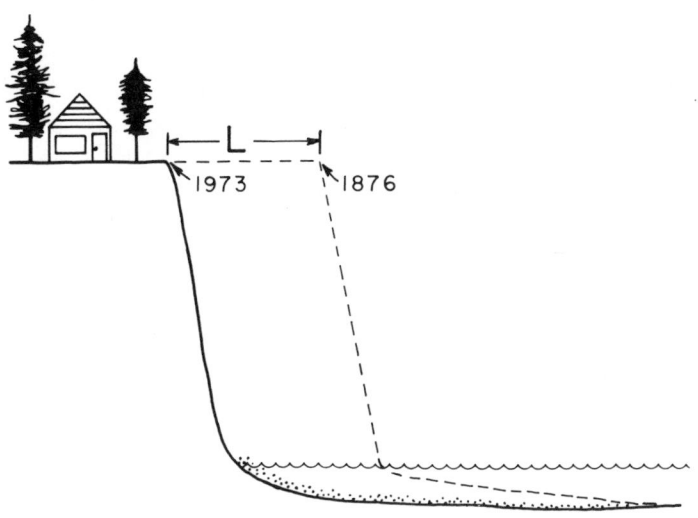

Cross section view

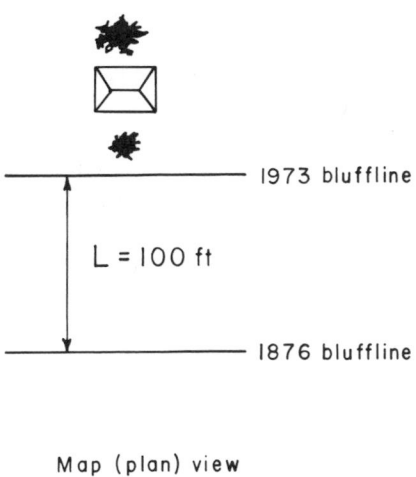

Map (plan) view

4.1   Determination of the long-term (average) erosion rate. The distance (L) is measured perpendicular to the two bluff lines. This distance is divided by the number of years between the bluff lines (T) to calculate the average erosion rate. In this example 100 feet per 97 years or about 1 foot per year.

tions of a mile) or stretches are evaluated in terms of: (1) the erosion rate, (2) the flood potential, (3) the shore material and relief, (4) beach width, and (5) the extent of shore protection structures. Note that erosion rates have been determined over a 100-year period and represent long-term averages (fig. 4.1). In fact, the rates vary greatly over periods of a few years, and are largely dependent upon lake level.

The maps and associated data on the following pages have been prepared to give you a thumbnail sketch of the different stretches along the shore. The patterns reflect the different shore materials that range from the barrier-marsh to rock. The hazard zones range from low risk (little chance of being flooded and/or a low erosion rate) to high risk (a good chance of being flooded and/or a high erosion rate). Wide beaches are beaches 30 feet or more in width, and discontinuous shore protection means that the stretch of shore in question is partially fronted by man-made structures. (No guarantee is given as to the effectiveness of the structures.)

Buyers, builders, or planners can assess the level of risk they are willing to take with respect to coastal hazards. Listing specific dangers, or conversely specific stability indicators, provides a basis for taking appropriate precautions in site selection, construction, or other land use. Our recommendation is to avoid high-risk zones, either completely or through significant setback onto stable land.

Small maps of large areas must be generalized; therefore, the buyer or planner must go beyond what we can show in this book and must evaluate each site individually. A relatively safe site can exist in a generally high-risk zone. Just as likely, a dangerous site can exist in a relatively low-risk zone.

Last, the best approach is to have a look at coastal property yourself, contact an expert for advice, and decide whether or not you can live happily with the risks.

In the rest of this chapter we first describe long lengths of shore (reaches) and then the relatively short lengths of shore (stretches) associated with each reach. In general, the stretches in a given reach have similar characteristics and are affected by similar processes. We do this to make the landowner or potential landowner aware of erosion and flood hazards. The subject order for each reach is: (1) physical nature and processes; (2) erosion and flood hazards; and (3) shore protection suggestions. The subject order for each stretch is: (1) nature of shore material, i.e. is it sand, clay, or rock? (2) height (relief) of the shore above the lake; (3) width and continuity of the beaches, i.e. are they continuous or discontinuous? (4) type, nature, and size of the man-made shore protection structures; and (5) erosion rates and flood hazards. Erosion rates are commonly determined over a 100-year period and thus represent a long-term rate (fig. 4.1). The rates, however, vary greatly over periods of a few years, and are largely dependent upon lake level.

Reach 1 (U.S.): Cherry Island, Michigan (Detroit River), to Huron, Ohio (Old Woman Creek)

This reach, at the west end of Lake Erie (fig. 1.1), is characterized by low relief (3 to 6 feet) and easily erodible clay banks or sand barriers, except for the rockbound shores of Catawba Island, Marblehead, and the Erie Islands. The shore is easily flooded because of the low relief (for example, the 1985 Palm Sunday and the December 1986 flood events). Because of its location, storm surges plague the shoreline residents during high lake level years. There is little sand in the beach-nearshore system except in the large pocket beaches between Catawba Island and Marblehead, off Port Clinton, and adjacent to the harbor structures from Port Clinton to Maumee Bay and south of Stony Point. As a result, significant erosion problems occur downdrift of the harbor structures. These harbor structures have starved the major bay mouth spits (Little Cedar Point and Woodtick Peninsula) at the mouth of the Maumee River. The spits and sand barriers are rapidly being lost to erosion because of an insufficient sand supply in the longshore transport system. Moreover, erosion of the shore has led to shore protection structures that have further reduced the amount of sand entering the system.

Major dikes, like those at Reno Beach and Point Place, offer good protection from flooding, but along much of the shore, smaller dikes and seawalls offer only poor to moderate protection against the floodwaters produced by major northeast wind storms. Structures built on ground level near the shore stand a good chance of being flooded. For this reason, those who choose to live on this lake floodplain should prepare to be flooded during periods of high lake level.

Erosion rates along this reach are the highest along the lake. Even along sheltered Sandusky Bay, erosion rates of 10 to 15 feet per year occur along unprotected segments of shore because the clay banks are so easily eroded. Erosion of the sand barriers is particularly unfortunate because the barriers constitute a rapidly disappearing ecological niche. Once the barriers are eroded, the marshes behind them will also be lost to the lake unless protected by dikes such as those at Pointe Mouillee and Little Cedar Point. The clay banks, which for the most part were part of the Great Black Swamp, are extremely fertile and contribute high yields of corn, wheat, and soybeans. These clay banks are also easily eroded.

If you must protect the shore, seawalls are an acceptable form of erosion prevention along this reach. Because of the relatively small waves, riprap seawalls will prevent most erosion, if they are maintained. A key point here is the prevention of overtopping, which can cause erosion from the back side of the structure. Rough-surfaced seawalls that slope gently to the lake dissipate the wave energy as it travels across the seawall surface, and are preferable to smooth-surfaced, vertical seawalls that deflect wave energy, leading to erosion on the back side of the structure as well as at the toe of the structure. Groins are a poor solution because of the sparse sand supply. Even if a groin were to trap sufficient sand to protect the shore on the updrift side, accelerated erosion would take place along a longer length of shore on the downdrift side. Breakwaters are a viable alternative to seawalls, but they are much more expensive to construct and maintain, and can have significant effects on the longshore transport of sand.

The rockbound shore, particularly at Catawba Island and the Bass Islands, is a special case. The rock is much more resistant to erosion than the sand barriers and clay banks, but it erodes nevertheless, mostly in the wave erosion zone at the toe of the bluff. Evidence of this can be seen in large (10 to 15 feet) out-of-place blocks that front the shore in places. The rock shore is undercut at lake level until the strength of the remaining rock is less than the force of gravity and the block collapses. This process is extremely irregular. The best way to ascertain whether the shore is stable is to get in a boat and have a look at the base of the bluff. If there is little undercutting (no more than 1 to 2 feet) the shore is probably stable, but if the undercutting is deeper there is a good chance that the bluff edge will collapse in the not-too-distant future. The prudent buyer will keep a safe distance from the shore.

Wayne and Monroe Counties, Michigan

Stretch 1: Cherry Island (mouth of Detroit River) to La Plaisance Creek (figs. 4.2 and 4.4). This stretch of shore is one of the most diverse in terms of land use along the Lake Erie shore. Uses range from recreation and wildlife areas of Sterling State Park and the Pointe Mouillee State Game Area, to residential areas such as Milleville Beach and Stony Point, to the coal-fired power plant at Monroe and the nuclear power plant (Enrico Fermi) north of Stony Point. Relief is extremely low, commonly less than five feet, along nearly the entire stretch. The small communities that front the shore are built on clay or fill land except for Stony Point, which occupies a rockbound shore composed of a dolomite breccia. This rock is much more resistant than the surrounding clay and sand-marsh, and as a result, projects out from the lakeshore. If there were no man-made shore protection in front of the more easily erodable deposits on either side, Stony Point could become another island. Sand barriers and the associated marshlands make up the majority of the shore, particularly between Milleville Beach and Estral Beach, Estral Beach and Stony Point, and Sterling Point and La Plaisance Creek (fig. 4.3).

The marshlands in the Pointe Mouillee area, now almost completely fronted by dikes, are relict deltaic deposits that formed at the mouth of the Huron River when the level of Lake Erie was lower. Beaches are narrow and discontinuous, mainly because of the scarcity of sand. Little sand-size sediment is transported to the lake by the tributary streams of this stretch, and the low clay banks contain little sand. As a result the sand in the barriers is rapidly being lost along the shore or washed over the barriers into the marshes. Moreover, the seawalls that now front most of the small communities along the shore deflect wave energy downward, eroding any sand at the base of the walls.

This shore is classified as a moderate risk zone. Erosion rates range from 0 to 5 feet per year. The highest rates occur along the barrier-marsh that fronts Sterling Beach State Park and the Pointe Mouillee area. The lowest rates are along the shores of older communities such as Estral Beach and Milleville Beach, where shore protection structures have existed for an appreciable length of time. The major problem, however, along this low-relief, moderate wave energy shore is not erosion, which can be reduced by well-built seawalls, but flooding. Major flood damage occurred here in the early 1950s and 1970s. The Corps of Engineers' Operation Foresight program provided about $4 million worth of flood protection in the early 1970s. The program is estimated to have prevented about $40 million in flood damage;

however, more recent storms have caused serious property damage. The area was hit hard by the 1985 Palm Sunday storm ($30 million in property damage) and a December 1986 storm during which easterly winds caused flooding that required the evacuation of about 25 homes and caused water damage to about 250 homes.

Flooding is a major problem because of the low relief, and because most of the houses are built at ground level. Unless an area is fronted by impermeable structures, the likelihood of flooding is great. More and more seawalls are being built along the shore, which may help; but if the seawalls are too low or are outflanked there will still be a flood problem. Homes built too close to the walls may be subject to wave damage. Home owners who want to remain in the floodplain can reduce flood damage by having their houses raised.

Stretch 2: La Plaisance Creek to Maumee Bay (Woodtick Peninsula) (fig. 4.4). This shore is quite similar to the Cherry Island-La Plaisance Creek stretch. The relief is generally less than five feet, and the shore consists of sand barriers, marsh, or clay banks. The developed segments of the shore, as at Bolles Harbor, North Shores, and Luna Pier (fig. 4.5), are built on the clay banks, whereas the Toledo Beach and Woodtick Peninsula segments are sand barriers and marsh. The trend of the spits in these areas indicates a net north to south sand transport direction. The narrow beaches, when present at all, indicate a scarce sand supply. This scarcity of sand, plus erosion by storm waves, has led to the dissection of Woodtick Peninsula into several

segments. If erosion continues, Woodtick Peninsula will be lost to the lake in several years. Both Woodtick and the Toledo Beach spit originated from headlands along which the sand built southward. As elsewhere, the seawalls, by deflecting wave energy downward, prevent the formation of beaches and further accelerate the longshore transport of sand, of which there is precious little.

The barrier-marsh areas are high risk zones, and the clay bank areas are moderate risk zones. Erosion rates range from 0 to 5 feet per year. Naturally, the rates are highest along the unprotected stretches of shore such as Woodtick Peninsula. Well-built seawalls can reduce erosion along these low relief stretches; however, even after seawalls are constructed the lake problem is not completely solved. Houses built along the shore will be flooded if not adequately protected from storm surges, which can reach several feet at this end of the lake. For example, under the Corps of Engineers' Operation Foresight program, about $2 million of flood protection structures were constructed. The protection is estimated to have prevented about $35 million in flood damage. However, lakeshore flooding, as along the Cherry Island-La Plaisance Creek stretch, will always be a threat along this shore.

Lucas County, Ohio

Stretch 3: North Maumee Bay to Maumee River (Maumee Bay) (fig. 4.4). The low (5 to 10 feet) relief and location at the west end of the lake make this stretch susceptible to lake flood-

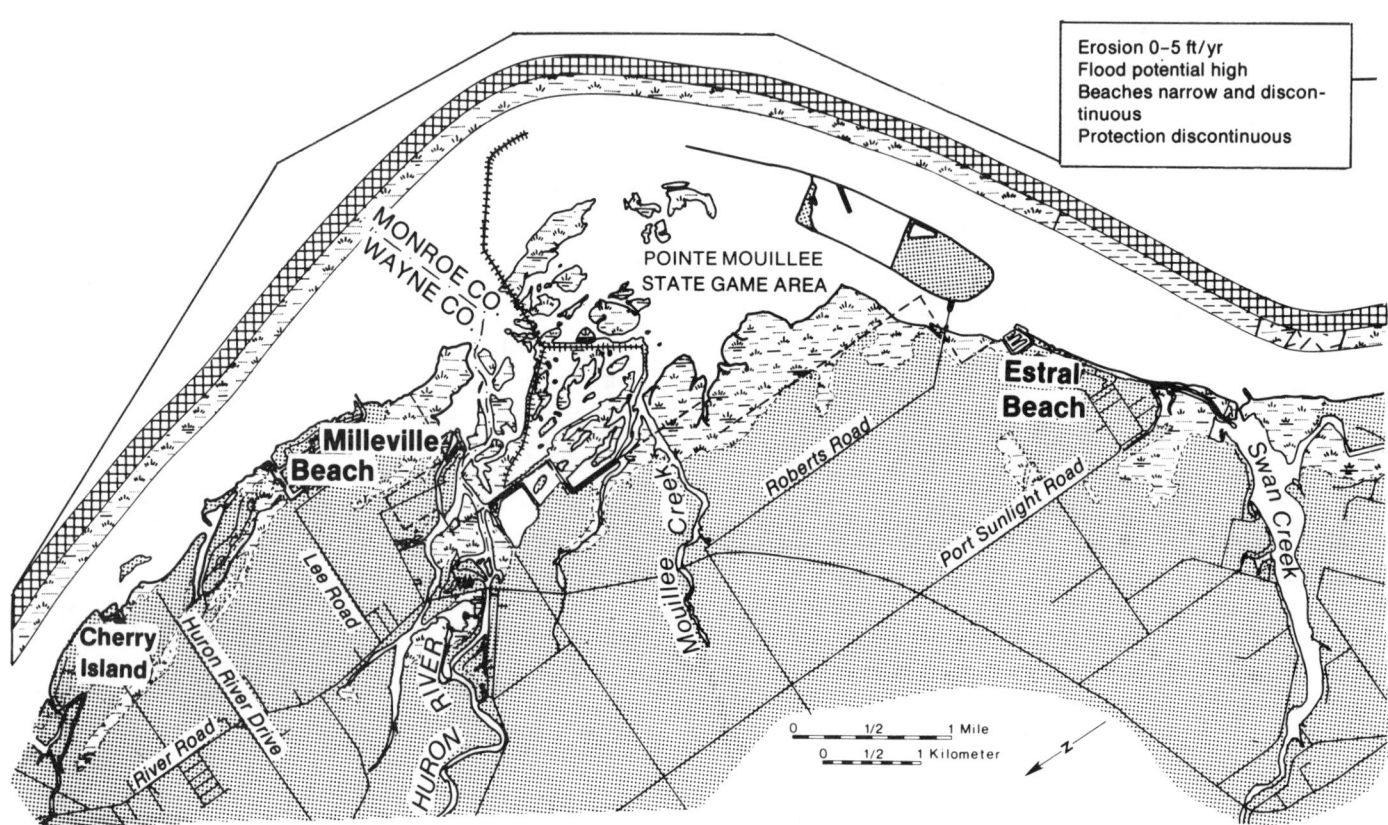

Erosion 0–5 ft/yr
Flood potential high
Beaches narrow and discontinuous
Protection discontinuous

4.2 Site analysis: Cherry Island to La Plaisance Creek

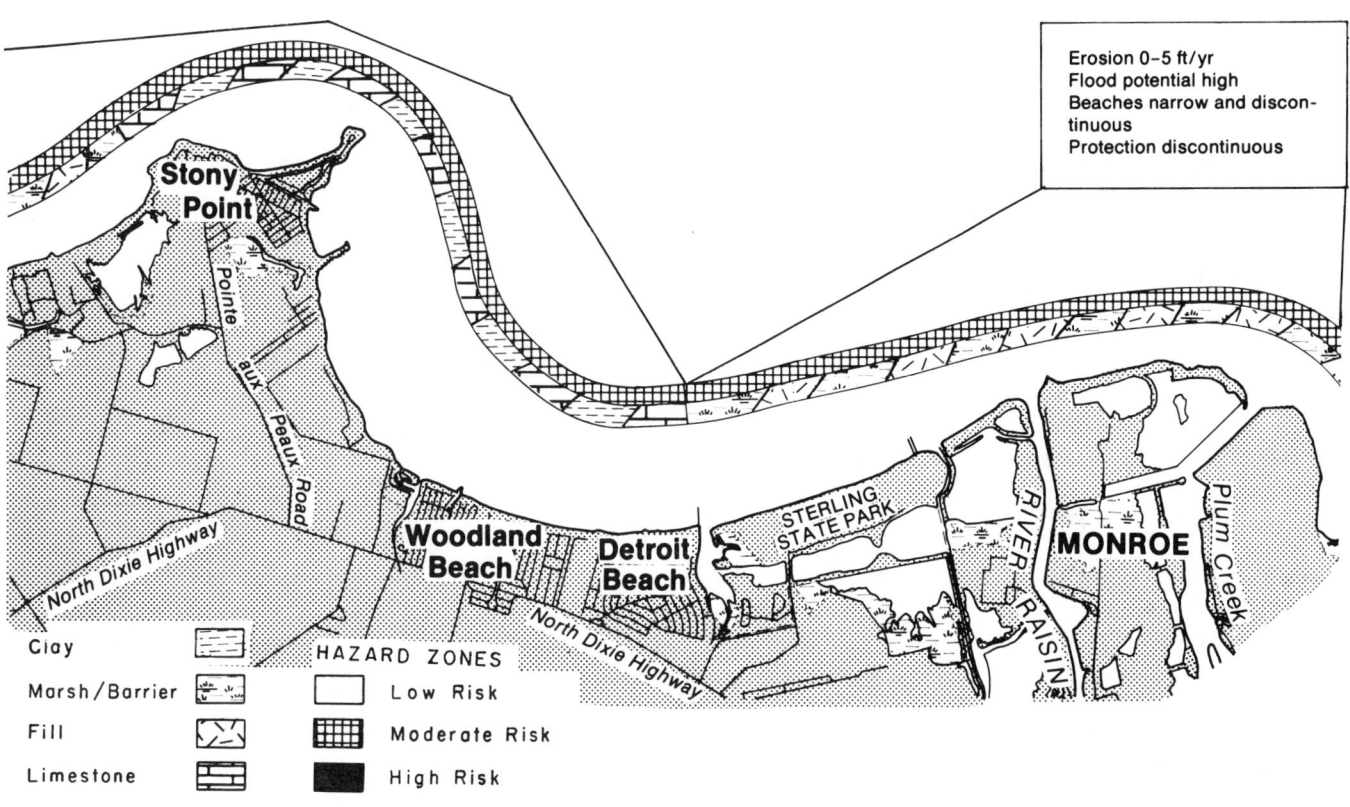

Erosion 0–5 ft/yr
Flood potential high
Beaches narrow and discontinuous
Protection discontinuous

Stony Point

Pointe aux Peaux Road

North Dixie Highway

Woodland Beach

Detroit Beach

North Dixie Highway

STERLING STATE PARK

RIVER RAISIN

MONROE

Plum Creek

Clay

Marsh/Barrier

Fill

Limestone

HAZARD ZONES

Low Risk

Moderate Risk

High Risk

4.3   Sand barrier and associated marshland, Estral Beach area. Low relief and easily eroded sand make barriers the most hazardous locations along a coast.

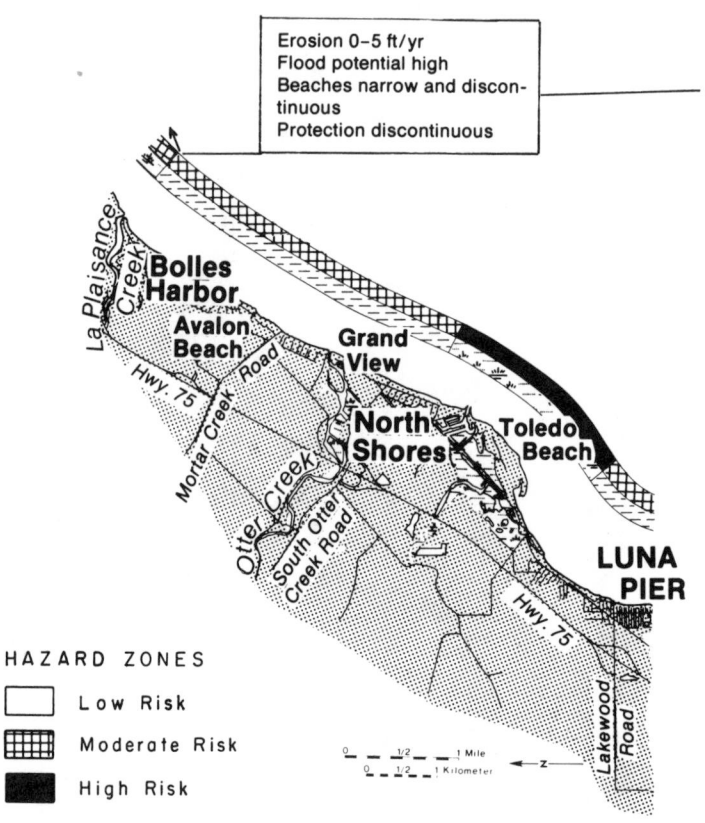

Erosion 0–5 ft/yr
Flood potential high
Beaches narrow and discontinuous
Protection discontinuous

HAZARD ZONES

☐ Low Risk

▦ Moderate Risk

■ High Risk

ing. Major flooding of developed areas in 1972–1973 led to the construction of an Operation Foresight dike at Point Place by the Corps of Engineers. This dike now fronts two miles of the stretch. The cost of the dike was about $2 million and the dike is estimated to have prevented about $3 million in flood damage in the early 1970s. The remainder of the stretch is also protected by structures (mostly seawalls) except for the Bay View Park area. Erosion is not a major problem because of the shore protection structures and the dikes, but if these structures are not maintained, rapid (10 to 15 feet per year) erosion will take place in

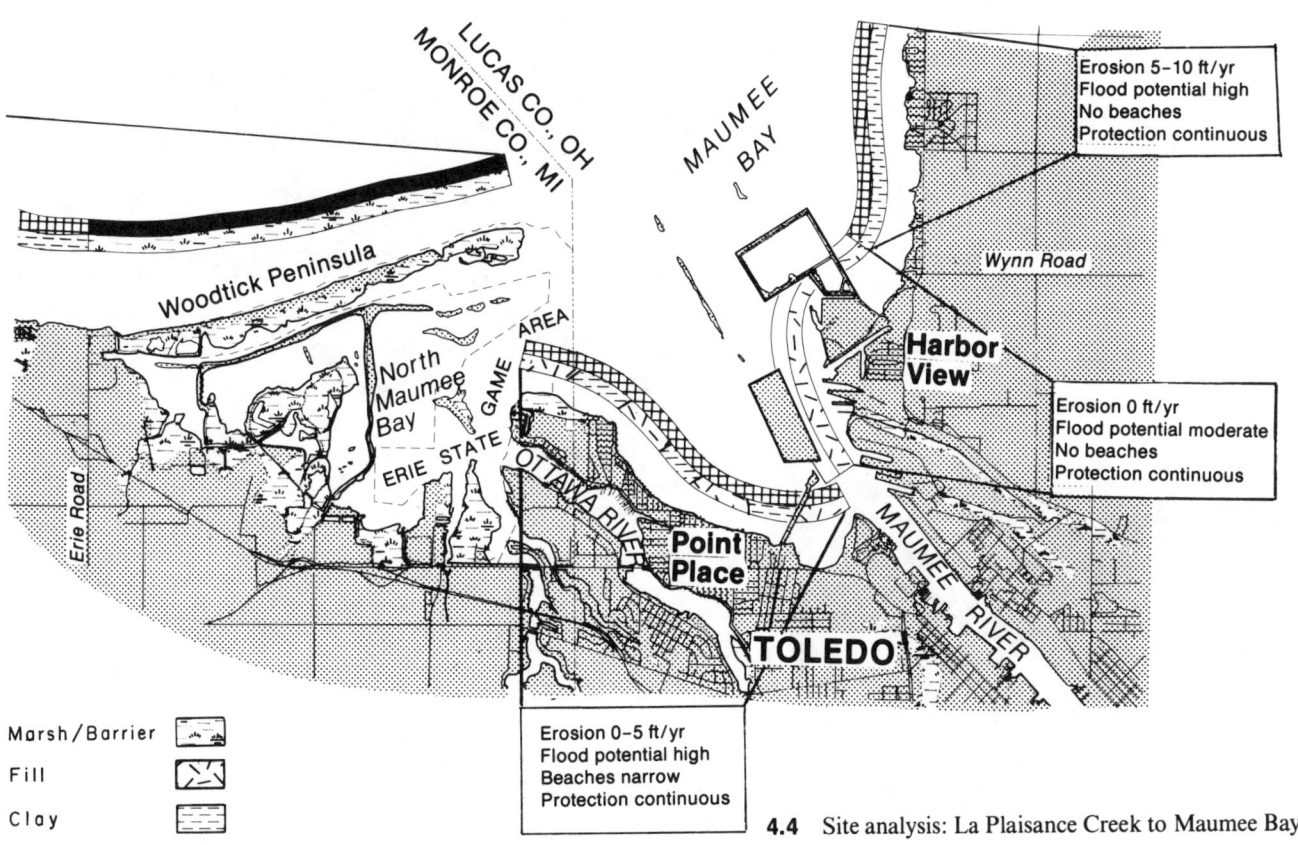

Erosion 5-10 ft/yr
Flood potential high
No beaches
Protection continuous

Erosion 0 ft/yr
Flood potential moderate
No beaches
Protection continuous

Erosion 0-5 ft/yr
Flood potential high
Beaches narrow
Protection continuous

Marsh/Barrier

Fill

Clay

4.4 Site analysis: La Plaisance Creek to Maumee Bay

the easily eroded lake clays and fill. The risk factor depends upon the continued maintenance of the structures (fig. 4.6).

Stretch 4: Maumee River to Wynn Road (Maumee Bay) (fig. 4.4). The low (5 to 10 feet) relief and western location make this stretch quite susceptible to lake flooding. However, unlike at the Toledo-Point Place stretch across the bay, the man-made dike disposal islands (polluted sediment from the Maumee River is deposited in these enclosures) that surround Harbor View block the storm waves and reduce the intensity of flooding. This stretch consists almost entirely of fill land, and erosion rates are therefore difficult to determine. Nonetheless the seawalls and dikes attest to erosion and flood problems along this industrial area.

4.5 Luna Pier. Low relief and location at the west end of the lake make development near the shore quite risky (see fig. 6.1).

4.6 Operation Foresight dike at Point Place. This dike provides protection from storm surges but has markedly changed the perspective of living along the lake. The dike, like all shore structures, needs periodic maintenance.

Stretch 5: Wynn Road to Niles Beach (Maumee Bay) (figs. 4.4 and 4.7). As along the other Maumee Bay stretches, the low relief of the clay banks and location at the western end of the lake make this stretch vulnerable to flooding. The low clay banks are sometimes protected by seawalls of heavy riprap, concrete, or sheet steel; where they are not, erosion rates can exceed 15 feet per year. From Wynn Road to Norden Road, and in the Niles Beach area, structural protection has reduced erosion markedly, but where the shore is unprotected, as at Maumee Bay State

Park, shore erosion rates are as high as anywhere along Lake Erie. This stretch was severely damaged by the storms in 1972 and 1973. The community of Niles Beach was abandoned because of extensive damage in the November 1972 storm (fig. 4.8) and is now part of Maumee Bay State Park. Flooding was the major problem, but waves caused damage too, particularly along the unprotected shore.

Stretch 6: Niles Beach to the tip of Little Cedar Point (Maumee Bay) (fig. 4.7). This stretch consists of marshland fronted by a riprap dike except for the shore between Niles Beach and Cousino Road. At the tip of Little Cedar Point (this is one of two Cedar Points along the Ohio shore of Lake Erie) sand deposited from the east-to-west longshore transport system extends to the west-northwest, offshore from the dike. Geologic studies between this point and Woodtick Peninsula across the bay indicate that the two points may have been connected at a lower lake level a few thousand years ago, forming a sand barrier across the bay. The increase in lake level and reduced sand supply, submerged and eroded the barrier leaving the two peninsulas as mute testimony to a former land bridge across Maumee Bay.

A riprap dike protects most of this stretch from erosion and flooding even during severe storms. The unprotected marsh, however, will recede rapidly and become an embayment of the lake unless it is fronted with some form of shore protection. Marshland is obviously not suitable or safe for development. Fortunately, most of this area is part of the Cedar Point National Wildlife Refuge.

Stretch 7: Little Cedar Point to Lakemont Landing (fig. 4.7). This protected wildlife refuge consists of marshland fronted by a riprap dike that extends about ten feet above mean lake level (fig. 4.9). Along the western two-thirds of the stretch the shoreline is relatively straight; along the eastern one-third, the shoreline forms an embayment that is backed by the dike. Remnants of the former spit that fronted Maumee Bay are preserved along the western one-third of the spit. Erosion rates of nearly 20 feet per year along short lengths of this stretch indicate the easily erodible nature of the sand. The combination of high lake levels and a reduced sand supply in the 1940s and 1950s (entrapment and deflection of sand transport by the jetty at Lakemont Landing) led to rapid erosion of the barrier and complete loss of the barrier along the eastern one-third of the stretch. Because of the diminished sand supply it is unlikely that the barrier beach will reestablish itself. At below-average lake levels beach nourishment along this stretch might well lead to formation of a new barrier beach. As along the portion of the point to the west, the riprap dike should protect the marsh behind it from both flooding and erosion. The eastern one-third of the stretch will continue to deepen as storm waves scour the bottom of the former marsh, but the dike built within the former marsh should protect the lowland on the landward side.

Stretch 8: Lakemont Landing to Wards Canal (fig. 4.7). This shore was part of the barrier beach-marsh system until the land was reclaimed in the late 1800s and early 1900s and an earthen dike was built for protection from the lake. Intense storms

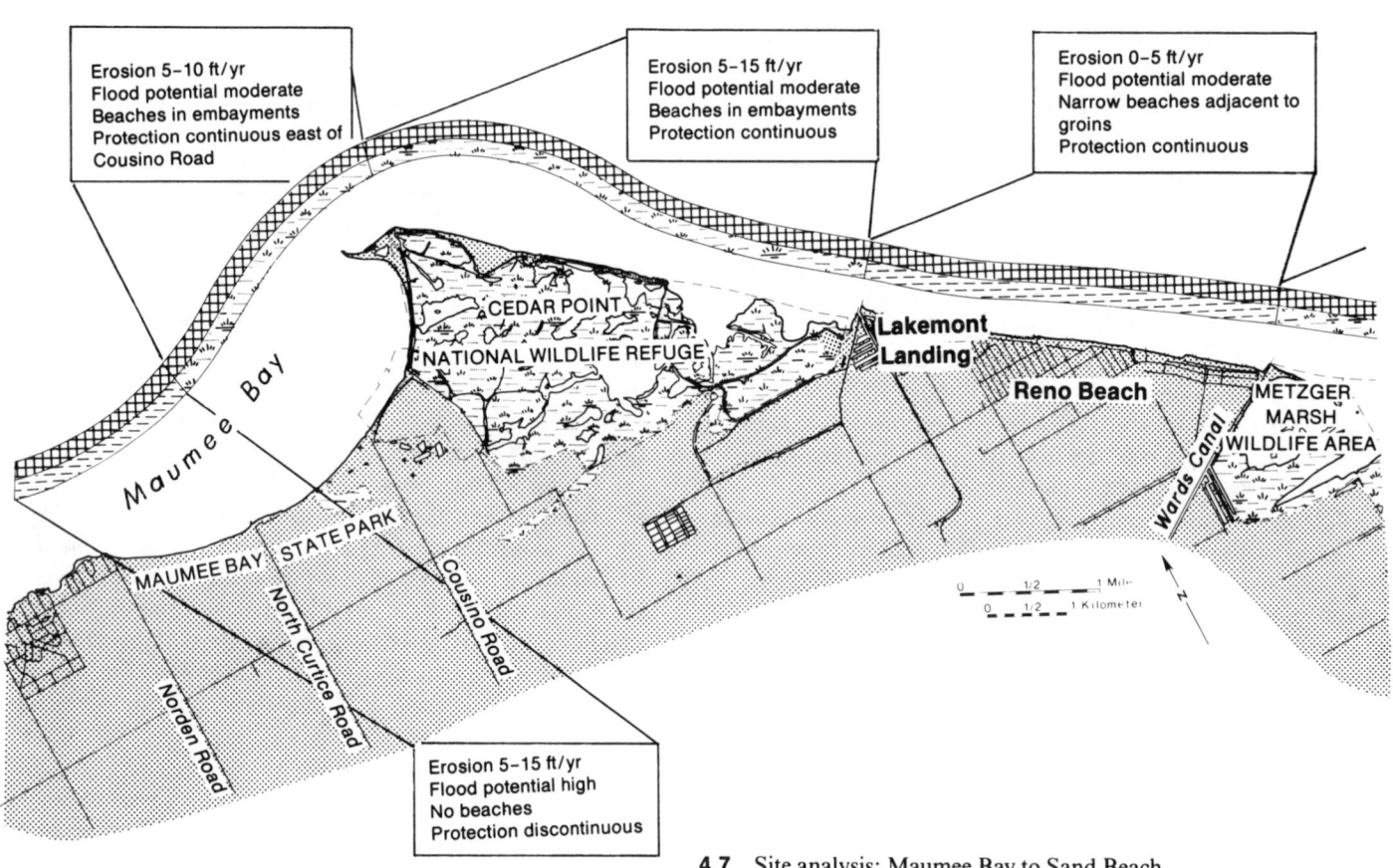

Erosion 5–10 ft/yr
Flood potential moderate
Beaches in embayments
Protection continuous east of
Cousino Road

Erosion 5–15 ft/yr
Flood potential moderate
Beaches in embayments
Protection continuous

Erosion 0–5 ft/yr
Flood potential moderate
Narrow beaches adjacent to
groins
Protection continuous

CEDAR POINT

NATIONAL WILDLIFE REFUGE

Lakemont
Landing

Reno Beach

METZGER
MARSH
WILDLIFE AREA

Maumee Bay

MAUMEE BAY STATE PARK

North Curtice Road

Cousino Road

Norden Road

Wards Canal

Erosion 5–15 ft/yr
Flood potential high
No beaches
Protection discontinuous

0   1/2   1 Mile
0   1/2   1 Kilometer

N

4.7   Site analysis: Maumee Bay to Sand Beach

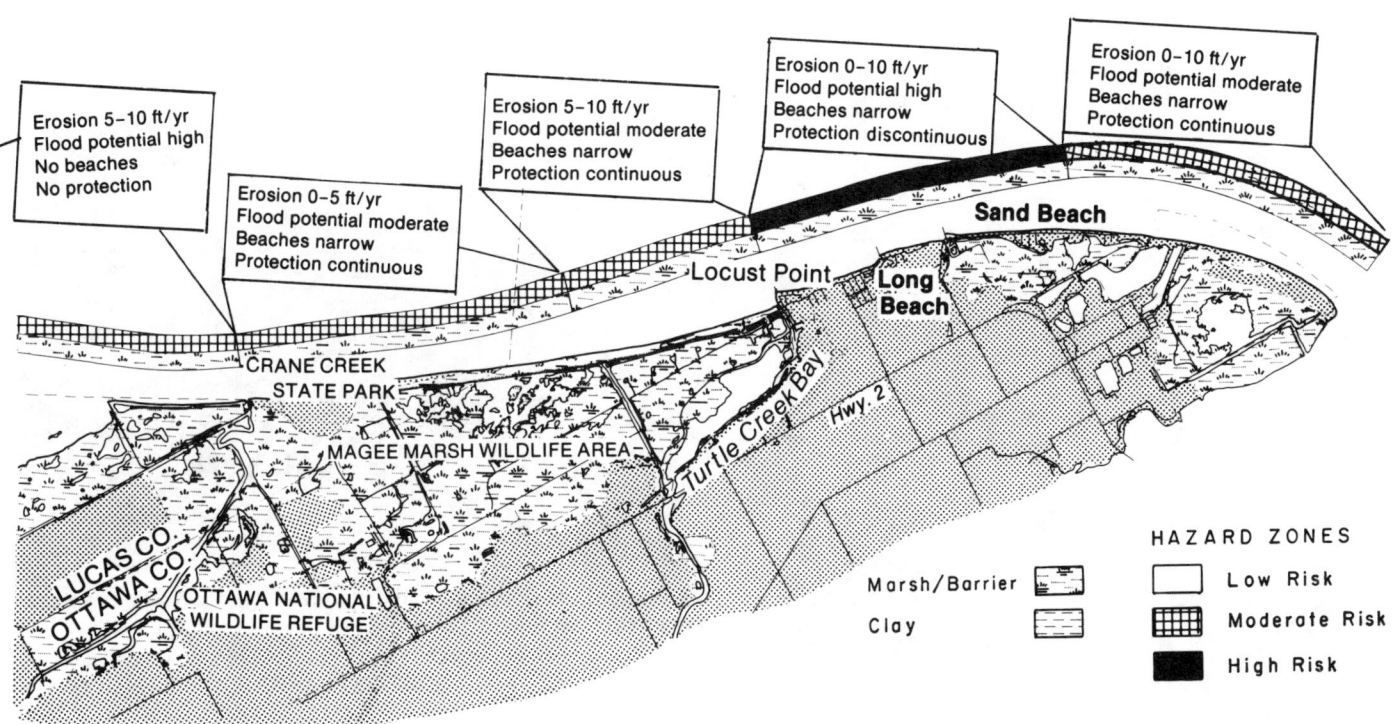

Erosion 5–10 ft/yr
Flood potential high
No beaches
No protection

Erosion 0–5 ft/yr
Flood potential moderate
Beaches narrow
Protection continuous

Erosion 5–10 ft/yr
Flood potential moderate
Beaches narrow
Protection continuous

Erosion 0–10 ft/yr
Flood potential high
Beaches narrow
Protection discontinuous

Erosion 0–10 ft/yr
Flood potential moderate
Beaches narrow
Protection continuous

Sand Beach

Locust Point

Long
Beach

CRANE CREEK
STATE PARK

MAGEE MARSH WILDLIFE AREA

Turtle Creek Bay

Hwy. 2

LUCAS CO.
OTTAWA CO.

OTTAWA NATIONAL
WILDLIFE REFUGE

HAZARD ZONES

Marsh/Barrier

Clay

Low Risk

Moderate Risk

High Risk

coupled with record lake levels in 1972–1973 badly damaged the earthen dike and associated structures such as groins. Many lakeshore homes built on the dike also were destroyed. This extensive damage led to the repair of the dike (Reno Beach-Howard Farms) under the Operation Foresight program (fig. 4.10). Repairs to the dike cost about $4 million and the estimated flood damage prevented during the high lake levels of the mid-1970s was $18 million.

4.8 Niles Beach after November 1972 storm. This community was badly damaged during the storms of 1972 and 1973 and has now been razed and incorporated into Maumee Bay State Park.

4.9 Cedar Point National Wildlife Refuge. Loss of sand to jetties and groins to the east (updrift), and high lake levels in the 1970s led to widespread erosion of the sand barrier fronting the marsh. The dikes were built to preserve the marsh for waterfowl.

4.10 Reno Beach-Howard Farms dike. This dike, which was badly damaged in the November 1972 storm, was rebuilt to protect houses on the lakeshore floodplain. This area is part of the Great Black Swamp that was drained by German settlers in the 1800s.

The problems experienced by landowners along this stretch epitomize the problems faced by all landowners along the low-relief western Lake Erie shore. The shore itself is a lake floodplain, subject to flooding from storm surges. Structures built adjacent to the lake on the floodplain are threatened by the flood hazard and are also subject to wave damage and potential destruction because of the low relief. The new riprap dike, which is a better shore protection structure than the earthen dike, probably will protect the property from wave attack and prevent flooding in all but the most severe storms. However, if the dike is not maintained and falls into disrepair the likelihood of future flooding and erosion will increase. Therefore this shore is considered one of moderate risk. The earthen core of the dike was failing in 1985; the Corps of Engineers hopes to rebuild the dike in the near future in cooperation with state and local governments (personal communication, Dick Bartz, Ohio Department of Natural Resources, 1985).

Lucas and Ottawa Counties, Ohio

Stretch 9: Wards Canal to Locust Point (fig. 4.7). This barrier beach and marsh zone is classified as high risk and appropriately has been conserved as wildlife refuges and parkland. Along the western two-fifths, the shore consists largely of marsh as the barrier beach has been eroded away. Along the eastern three-fifths a riprap-covered dike forms the barrier. The exception is a short length of barrier beach at Crane Creek State Park where a jetty and groins have trapped sand and kept the barrier intact. The barrier beach-marsh system is owned by the state and federal governments that manage it for waterfowl. At Crane Creek, there is a lakeshore park and a nature center with an emphasis on migratory waterfowl.

This stretch is typical of the Michigan and Ohio shore at the west end of the lake with its sandy barriers backed by marshes. Because the sand is so easily eroded, the jetties constructed at Locust Point, Crane Creek, and west of Crane Creek have helped stabilize the shore to the east of the structures by trapping sand from the longshore transport system, but on the west (downdrift) side of the structures the loss of sand has resulted in erosion of the barrier and encroachment of the lake into the marsh. The dikes should prevent erosion of the shore if properly maintained; but because of the sparse sand supply and the jetties, it is unlikely that the barriers will ever reestablish themselves in front of the embayments unless there is a major, man-made beach nourishment project.

Ottawa County, Ohio

Stretch 10: Bass Islands (fig. 4.11). The Bass Islands (South, Middle, and North) are made up of Silurian dolomite (a carbonate rock) capped in most places by Pleistocene glacial till. Beaches are narrow and discontinuous, and are commonly present as small pockets between enclosing rocky headlands. For unknown reasons, the dolomite making up the islands eroded more slowly than the adjacent rock, leaving these picturesque geologic markers in the middle of Lake Erie. The dolomite making up the low bluffs of South Bass is eroded by storm waves near lake level, and it unpredictably collapses. The dolomite or till making up the banks at Middle and North Bass is sufficiently low that the storm waves erode the shore directly back, but when the bank consists of till overlying dolomite, more rapid erosion of the till will commonly leave a bench of the more resistant dolomite in front of the less resistant till. The islands are known for their beauty, vineyards and wineries, and on South Bass (fig. 4.12) Perry's Monument and Ohio State University's Stone Limnological Laboratory.

Average erosion rates range from about 0.5 feet per year along the rockbound South Bass shore to about 2 feet per year along the mostly till shore of North Bass. Naturally the rates are highest along the unprotected shores composed of till on all the islands. Flooding is a problem, particularly along the lower relief shores of Middle and North Bass during northeast storms at times of high lake level. However, because of the largely undeveloped nature of the islands (particularly Middle and North Bass) the problems associated with erosion and flooding are minor in comparison to that of highly developed mainland stretches.

Stretch 11: Locust Point to Toussaint River (figs. 4.7 and 4.13). This high risk stretch, including the small communities of Locust Point, Long Beach, and Sand Beach, is one of the most exposed, and thus most vulnerable areas along the western basin shore. Even though there is nearly continuous protection in the

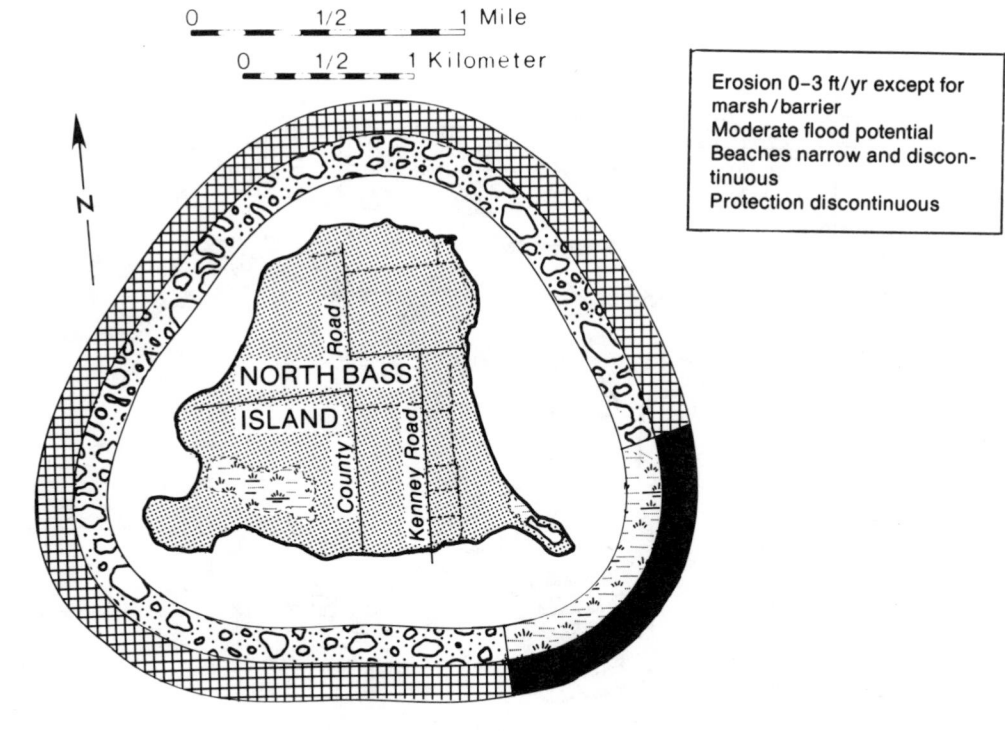

0   1/2   1 Mile

0   1/2   1 Kilometer

N

NORTH BASS
ISLAND

County Road

Kenney Road

Road

Erosion 0-3 ft/yr except for
marsh/barrier
Moderate flood potential
Beaches narrow and discon-
tinuous
Protection discontinuous

Glacial Drift

Marsh/Barrier

HAZARD ZONES

Low Risk

Moderate Risk

High Risk

4.11 Site analysis: Bass Islands

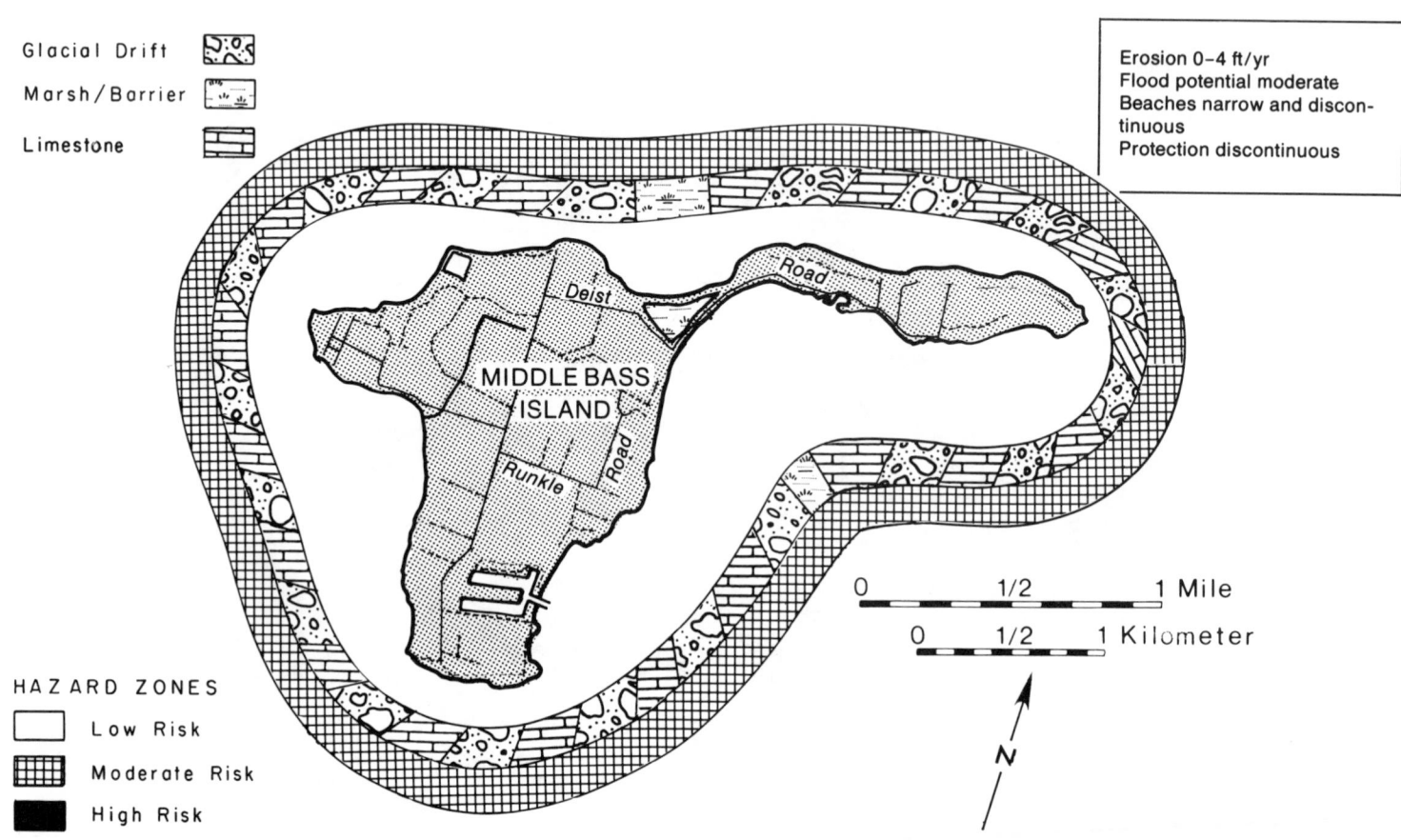

Glacial Drift

Marsh/Barrier

Limestone

Erosion 0–4 ft/yr
Flood potential moderate
Beaches narrow and discontinuous
Protection discontinuous

MIDDLE BASS ISLAND

Deist   Road

Runkle   Road

0   1/2   1 Mile

0   1/2   1 Kilometer

N

HAZARD ZONES

Low Risk

Moderate Risk

High Risk

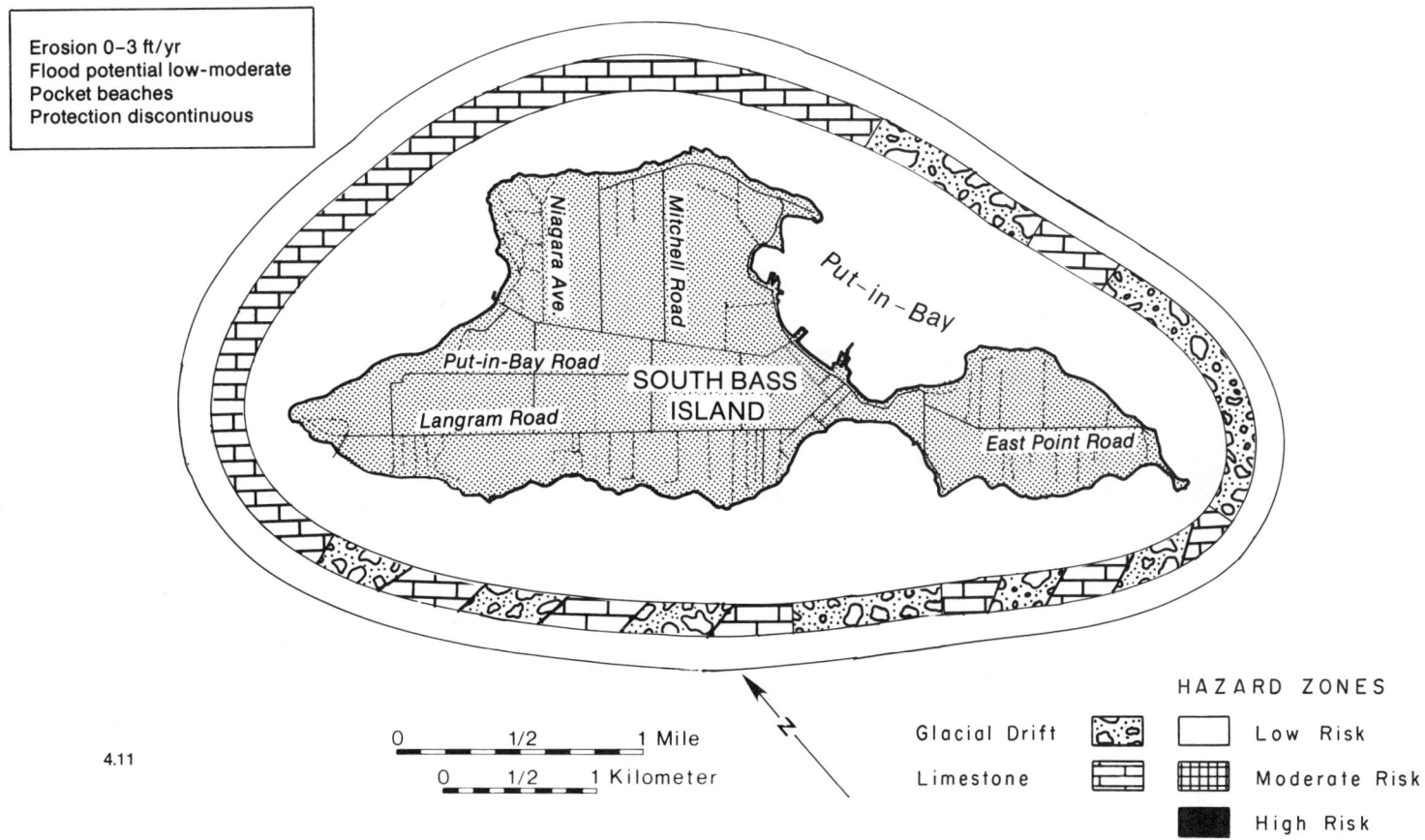

Erosion 0–3 ft/yr
Flood potential low-moderate
Pocket beaches
Protection discontinuous

Put-in-Bay

Niagara Ave

Mitchell Road

Put-in-Bay Road

SOUTH BASS
ISLAND

Langram Road

East Point Road

4.11

0   1/2   1 Mile

0   1/2   1 Kilometer

N

HAZARD ZONES

Glacial Drift

Limestone

Low Risk

Moderate Risk

High Risk

4.12   The southern end of South Bass Island. The more resistant rock of the islands has contributed to their preservation. However, they are slowly eroding.

form of seawalls and groins fronting the lakeshore communities along the western one-half of this stretch, the low (less than five feet) relief, sand shore, and the proximity of the homes to the lake make flood and erosion damage quite likely (fig. 4.14). For example, in 1972 and 1973, about 30 homes in the Sand Beach area were badly damaged during storms because the homes were located directly on the beach. Even the homes inland from the beach were damaged by flooding, and the inhabitants were inconvenienced by sand washed over and through the dunes onto roads and yards. Erosion is quite variable along this barrier-marsh shore. Erosion rates are as high as eight feet per year. In two areas, the east side of the Locust Point jetty and the east side of Sand Beach, the shoreline has built lakeward. These areas, however, are now eroding and are likely to continue eroding in the future. The shore protection fronting Locust Point, Long Beach, and Sand Beach will help to retard shore erosion, but the communities (and especially the homes along the shoreline) will still be risk prone, subject to wave erosion and flooding. The riprap dike that fronts most of the Davis Besse Nuclear Power Plant property to the east of Sand Beach should protect the barrier and the marsh from erosion. Nonetheless, all such structures must be maintained at a cumulative cost that may approach the original property value.

Stretch 12: Toussaint River to Rock Ledge (fig. 4.13). This stretch, although similar to the Locust Point-Toussaint River stretch, is more sheltered from easterly waves because of the peninsula of Catawba Island to the east. Because of the low relief (usually less than five feet), and easily eroded sand barriers, major erosion and flooding should be expected. Such was the case during the 1972 and 1973 storms (fig. 4.15A). In spite of the risk, the shore continues to be developed outside of the privately owned marsh and the Ottawa National Wildlife Refuge (fig. 4.15B). Most of the developed land is fronted with shore protection structures that will retard erosion, but flooding remains a major problem. Moreover, if the structures are not maintained, erosion will again be a problem. Homes built directly on the shoreline invite trouble.

Erosion rates are not as variable as along the stretch to the west. The long-term rates are 0 to 5 feet per year, with a general decrease in erosion near Port Clinton where there are more man-made structures. The Portage River area is the most stable. Jetties at the river mouth have trapped sand from the longshore transport system, building beaches on either side of the river, and the natural embayment between the Toussaint River and Rock Ledge has led to appreciable accumulation of sand, particularly at Port Clinton.

Stretch 13: Rock Ledge to Gem Beach (Catawba Island) (fig. 4.16). This low risk rocky headland extends out into the lake because its rock is more resistant than those that formerly flanked it to the east and west. Catawba Island, as it is known, is made up of Silurian carbonate rocks that dip gently to the southeast. The result is a resistant, 10-to-15-foot-high, rockbound shore that is much less susceptible to erosion and flooding than the sandy, low relief shore to the west. Caution is advised, however,

because occasional rock falls occur when the cliffs are undercut. The low-lying areas along the east side of Catawba Island are flooded occasionally by major lake storms. Much of this shore is built up, or will be developed, as the island area is one of the more attractive areas of the entire lake.

Shore protection structures are built along the lower-relief areas of the shore, particularly where the more easily erodible till overlies the rock. Erosion rates are low, generally 0 to 2 feet per year. Occasional cliff failures make it advisable for the property owner to examine the base of his cliff to see if it is undercut and thus likely to fail. Where the shore is less than five feet above the long-term average lake level, one can expect flooding.

Stretch 14: Gem Beach to Lakeside (fig. 4.16). This barrier beach connects the carbonate headlands of Catawba Island and Marblehead. The barrier likely formed as a bay-mouth bar when the ancestral Portage River flowed between Catawba Island and Marblehead at a lower lake level. When lake level rose, the shore was eroded back to the river at Port Clinton, and the river had a shorter path to the lake.

The low relief (five feet) and sand imply an easily erodible shore, but surprisingly the shore has remained quite stable for the last 100 years. There have, however, been significant man-made changes. The barrier behind the East Harbor segment has been augmented by dredge spoil, increasing its width by two to three times, and the East Harbor segment, as well as part of the Middle Harbor segment, has been fronted by a dimension stone seawall since 1953. This seawall was badly damaged during the

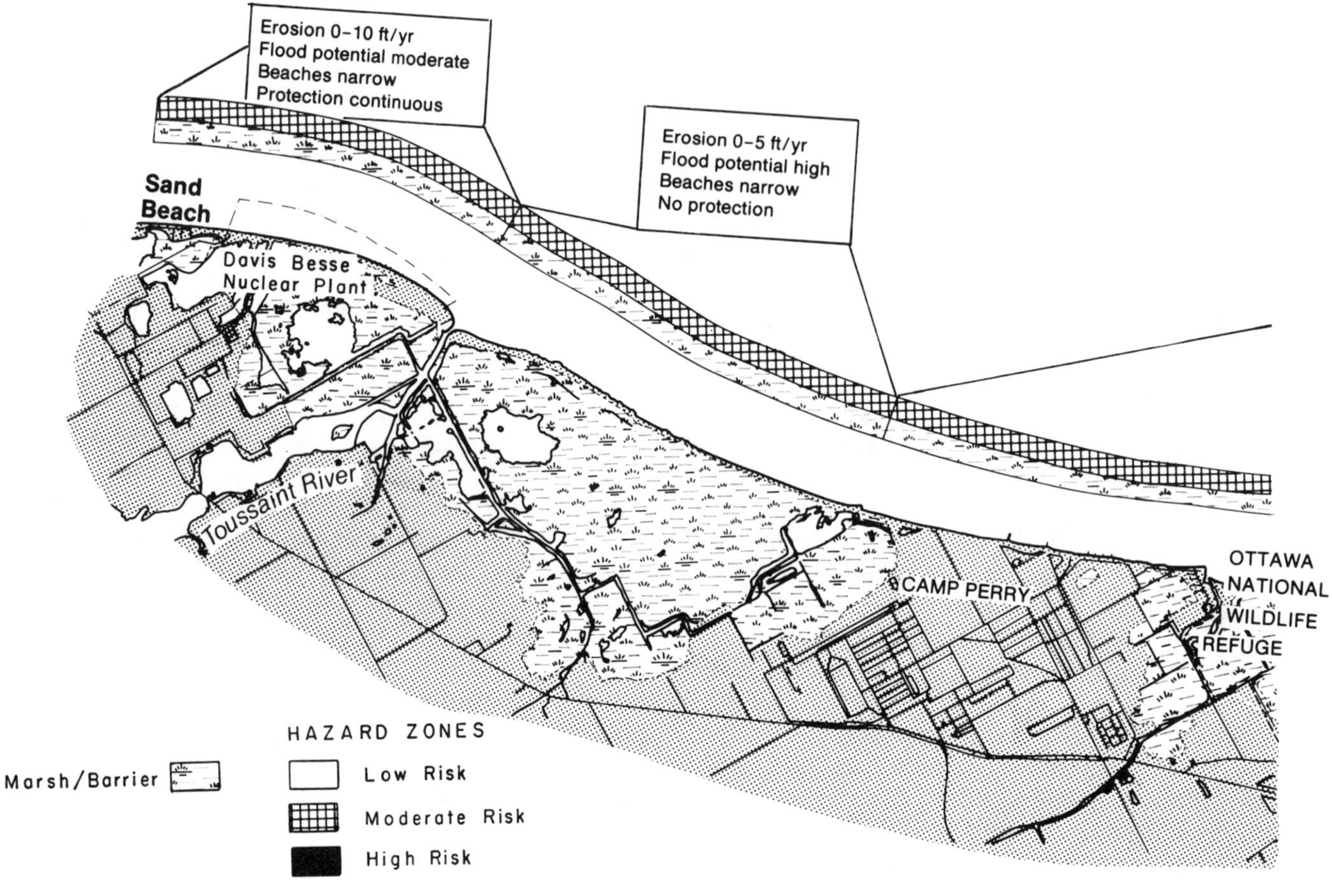

Erosion 0-10 ft/yr
Flood potential moderate
Beaches narrow
Protection continuous

Erosion 0-5 ft/yr
Flood potential high
Beaches narrow
No protection

Sand Beach

Davis Besse Nuclear Plant

Toussaint River

CAMP PERRY

OTTAWA NATIONAL WILDLIFE REFUGE

HAZARD ZONES

Marsh/Barrier

Low Risk

Moderate Risk

High Risk

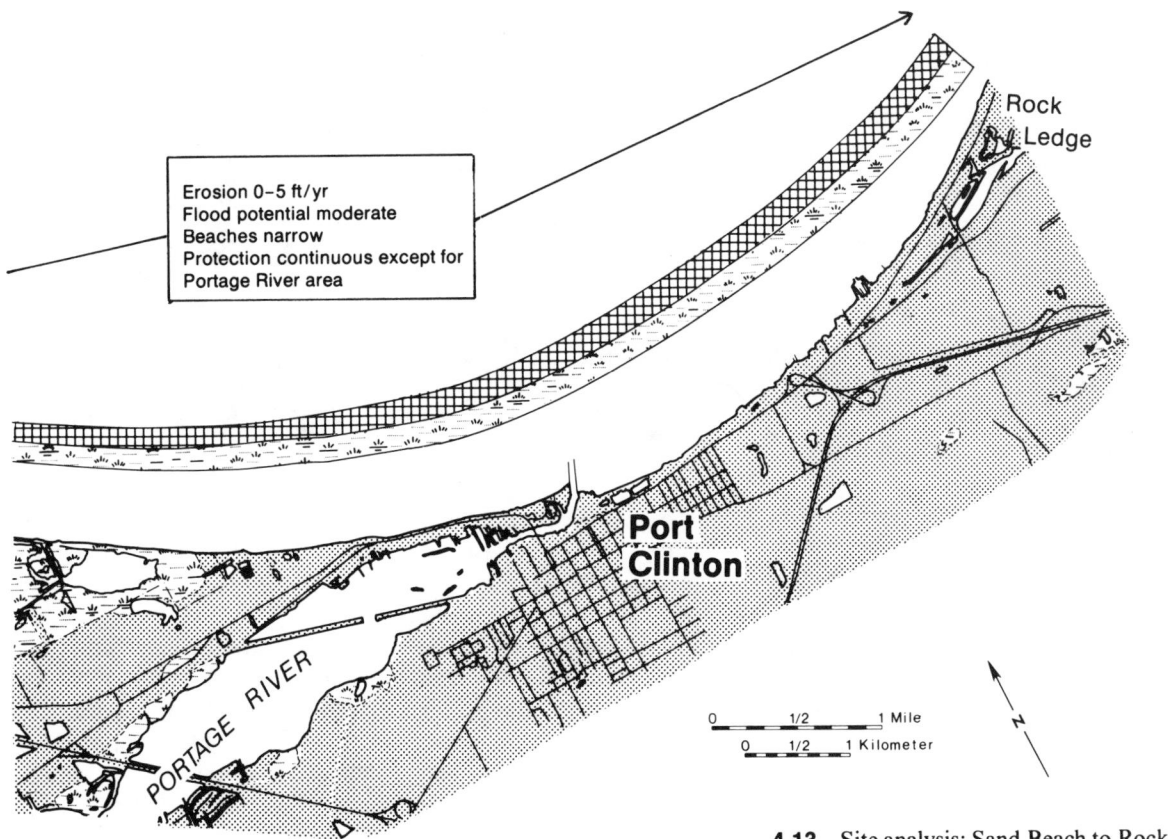

Erosion 0–5 ft/yr
Flood potential moderate
Beaches narrow
Protection continuous except for
Portage River area

Rock
Ledge

**Port
Clinton**

PORTAGE RIVER

0   1/2   1 Mile

0   1/2   1 Kilometer

N

4.13   Site analysis: Sand Beach to Rock Ledge

4.14 Sand Beach. Homes built on lake side of sand barrier, a vulnerable location for both flooding and erosion. (A) With seawall and no beach. (B) No seawall with narrow beach. (C) Waves breaking on front door at Sand Beach, spring 1973. Photo courtesy of the *Toledo Blade*.

beach. Along the unprotected segment, erosion of the barrier has storms and high lake levels of 1972 and 1973. The seawall was overtopped and sand was eroded from the back side, leading to erosion of the riprap core and collapse of the stone slabs. The effect of the seawall on erosion and flooding are uncertain. However, the seawall reflects wave energy causing wave scour at the base of the seawall and restricting the development of a natural

been more severe, but the beach is building back up because wave energy is more uniformly distributed along the gently sloping, barred nearshore.

Erosion rates are low, on the order of 1 to 2 feet per year. The most vulnerable segment is the unprotected sand barrier that fronts Middle Harbor. Moreover, the abundant sand supply along this stretch, as evidenced by the gentle, sandy, nearshore slope and the multiple nearshore bars, make it likely that the barrier will rebuild itself after high lake level erosion. The entire stretch is subject to flooding because of its low relief and its exposed location at the west end of the lake.

4.15 (A) Port Clinton after November 1972 storm. A lakeshore flood-plain. (B) Port Clinton shore, 1985. Development continues in the coastal zone.

Stretch 15: Lakeside to Bay Point (Marblehead) (figs. 4.16 and 4.17A). The resistant Columbus limestone forms this head-land, jutting into the lake on the north side of Sandusky Bay. Because of the site's importance to navigation, a lighthouse was built at Marblehead in 1821, three years after the Erie, Pennsylvania, and Buffalo, New York, lighthouses. This lighthouse is now the oldest operating lighthouse on the Great Lakes. In general, Marblehead is fairly stable; the rock erodes slowly. Problems do occur along the east side of the point, where the combination of low relief (less than ten feet) and the local presence of till makes the shore vulnerable to erosion and flooding during major northeast wind storms. Bay Point (fig. 4.17A), along the southwest side of the headland, is a sand spit (bay mouth) that partly fronts Sandusky Bay. This spit probably formed when Cedar Point (to the southeast) built across the bay. There is little, if any, sand along the Marblehead shore to be transported southward to build such a spit.

Overall, the shore is moderately developed, and is fronted here and there with shore erosion structures, particularly along the north-facing shore. In addition, man-made fill was added to the lake in several places to make small boat slips and harbors. The scarcity of sand combined with wave reflection results in a scarcity of beaches (except pocket beaches) along the stretch. Erosion rates are low, usually about one foot per year along the rockbound Marblehead shore, whereas Bay Point has experienced erosion of as much as four feet per year at its northern end. Naturally, Bay Point is the most unstable segment of shore because of its easily erodible sand. The relatively low relief along the east-facing segment makes it subject to flooding. In the early 1970s Ohio Route 163 was closed in places because of floodwaters and riprap that had been carried onto the roadway by storm waves.

Stretch 16: Bay Point to Muddy Creek (north shore of Sandusky Bay) (fig. 4.17A). Sandusky Bay is relatively well protected from open lake waves by the Bay Point and Cedar Point spits, which act as barriers to the large northeast storm waves. However, during major storms lake level can be set up 3 to 4 feet in the bay, which at high lake levels causes extensive flooding along the low relief, flat-lying Sandusky Bay shore. In general, the till shore between Bay Point and Danbury is more resistant to erosion than the lake-clay shore between Danbury and Muddy Creek, but because the western part of the bay is more exposed than the eastern part, there is not that much difference in erosion rates. Because of the relatively low energy setting, seawalls can provide effective shore protection. The scarcity of sand precludes the use of groins. Overall, the shore is sparsely to moderately built up. Development has not been rapid because of the shallow, muddy bottom and lack of sand beaches, but as land becomes scarcer along the open lakeshore there will be more pressure to develop this area.

Erosion rates range from 0 to 5 feet per year. The zero rates occur along segments with man-made protection. Besides erosion, which can be combatted with relatively low-cost structural protection, the entire stretch, particularly the segment west of

Danbury, is subject to flooding. Houses built on stilts might be appropriate in this area.

Sandusky and Erie Counties, Ohio

Stretch 17: Muddy Creek to Cedar Point (south shore of Sandusky Bay) (figs. 4.17B and 4.18). The setting and processes along the south side of the bay are quite similar to those along the north side. In general, the Upper (western) Bay is characterized by low clay banks that can erode at prodigious rates (about 15 feet per year) whereas the Lower (eastern) Bay is characterized by man-made structures in the form of dikes and seawalls. In the early 1970s, after severe flooding at Muddy Creek, Whites Landing, Crystal Rock, and Bay View, the U.S. Army Corps of Engineers constructed flood control dikes at these communities under the Operation Foresight program. In 1985 the Corps of Engineers rebuilt the dikes at Whites Landing and Crystal Rock. The cost of constructing the dikes was about $300,000. The flood damage prevented by storm surges in the early 1970s was estimated at about $2 million. In addition to protecting the shore from flooding, these dikes also protect the shore from erosion. Because of the low relief and low energy setting, as along the north shore of the bay, seawalls can provide effective shore protection.

Erosion rates range from 0 to 15 feet per year along the Upper Bay shore to 0 to 5 feet per year along the Lower Bay shore. The rates are zero where the shore is fronted by well-built sea-

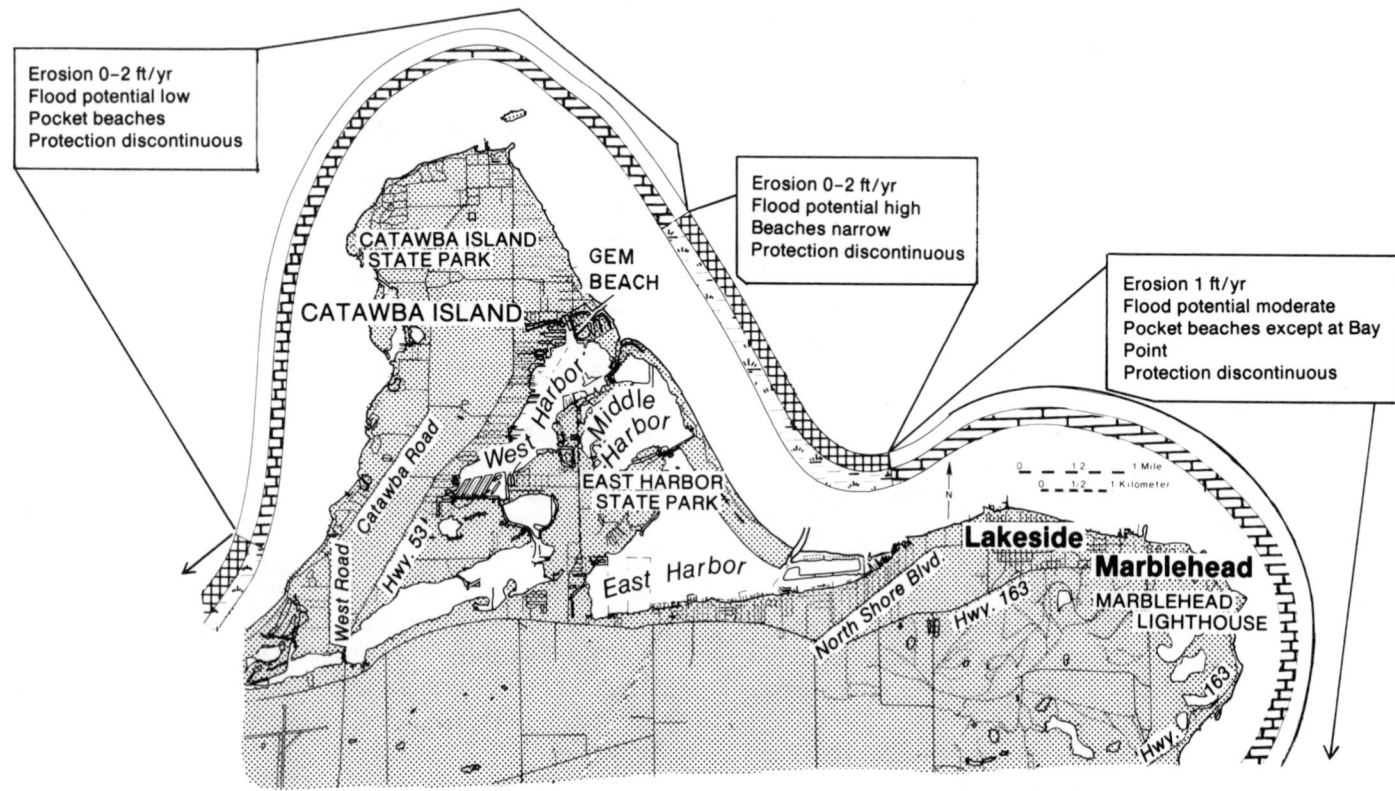

Erosion 0–2 ft/yr
Flood potential low
Pocket beaches
Protection discontinuous

Erosion 0–2 ft/yr
Flood potential high
Beaches narrow
Protection discontinuous

Erosion 1 ft/yr
Flood potential moderate
Pocket beaches except at Bay Point
Protection discontinuous

CATAWBA ISLAND STATE PARK

GEM BEACH

CATAWBA ISLAND

West Harbor

Middle Harbor

EAST HARBOR STATE PARK

East Harbor

West Road

Catawba Road

Hwy. 53

North Shore Blvd

Lakeside

Marblehead

Hwy. 163

MARBLEHEAD LIGHTHOUSE

Hwy. 163

0   1/2   1 Mile
0   1/2   1 Kilometer

4.16   Site analysis: Rock Ledge to Bay Point, and Kelleys Island

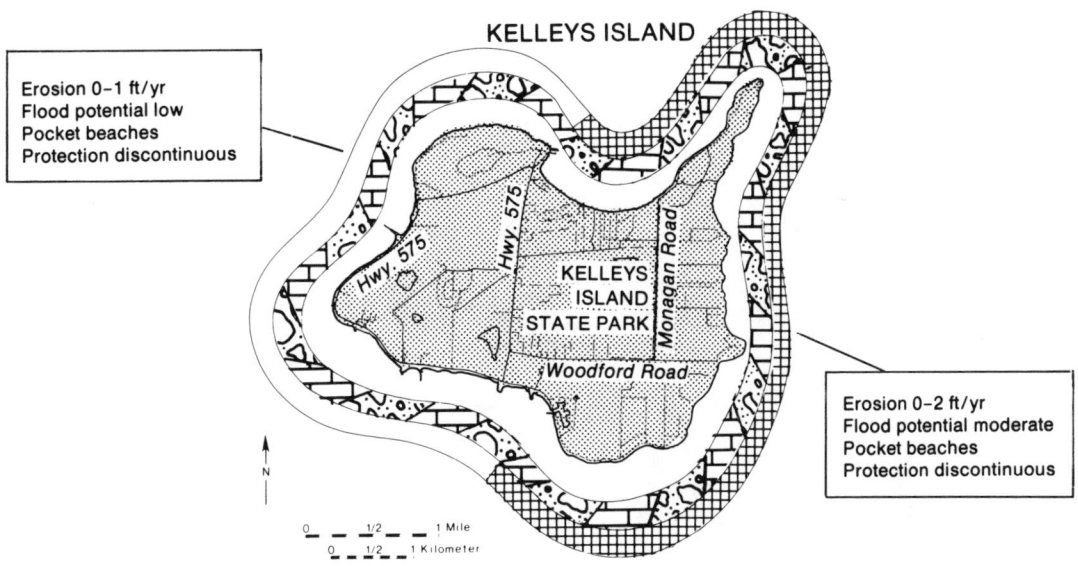

KELLEYS ISLAND

Erosion 0–1 ft/yr
Flood potential low
Pocket beaches
Protection discontinuous

Hwy. 575

Hwy. 575

Monagan Road

KELLEYS
ISLAND
STATE PARK

Woodford Road

Erosion 0–2 ft/yr
Flood potential moderate
Pocket beaches
Protection discontinuous

N

0 ___ 1/2 ___ 1 Mile
0 ___ 1/2 ___ 1 Kilometer

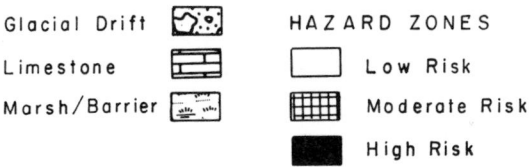

Glacial Drift

Limestone

Marsh/Barrier

HAZARD ZONES

Low Risk

Moderate Risk

High Risk

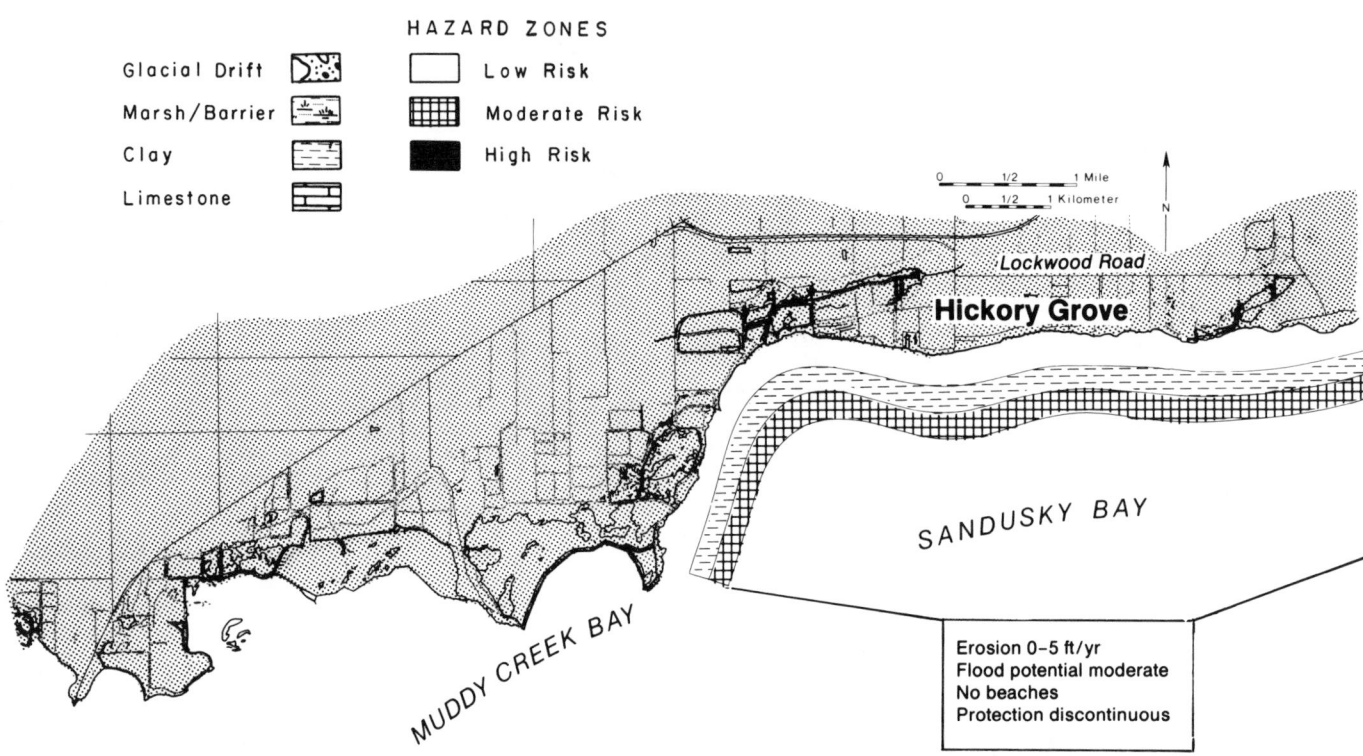

HAZARD ZONES

Glacial Drift

Marsh/Barrier

Clay

Limestone

Low Risk

Moderate Risk

High Risk

0   1/2   1 Mile
0   1/2   1 Kilometer

N

Lockwood Road

Hickory Grove

SANDUSKY BAY

MUDDY CREEK BAY

Erosion 0-5 ft/yr
Flood potential moderate
No beaches
Protection discontinuous

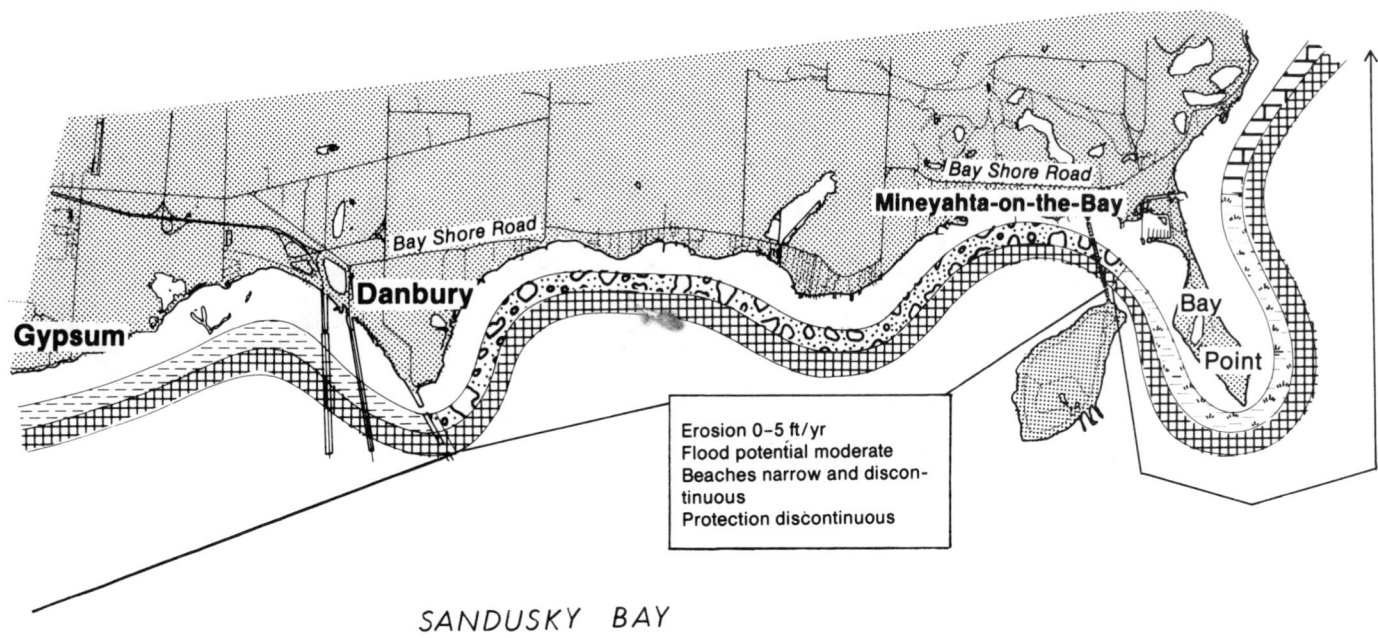

Gypsum

Danbury

Bay Shore Road

Mineyahta-on-the-Bay

Bay Shore Road

Bay
Point

Erosion 0–5 ft/yr
Flood potential moderate
Beaches narrow and discon-
tinuous
Protection discontinuous

SANDUSKY BAY

4.17A   Site analysis: Sandusky Bay. North shore, Bay Point to
Muddy Creek.

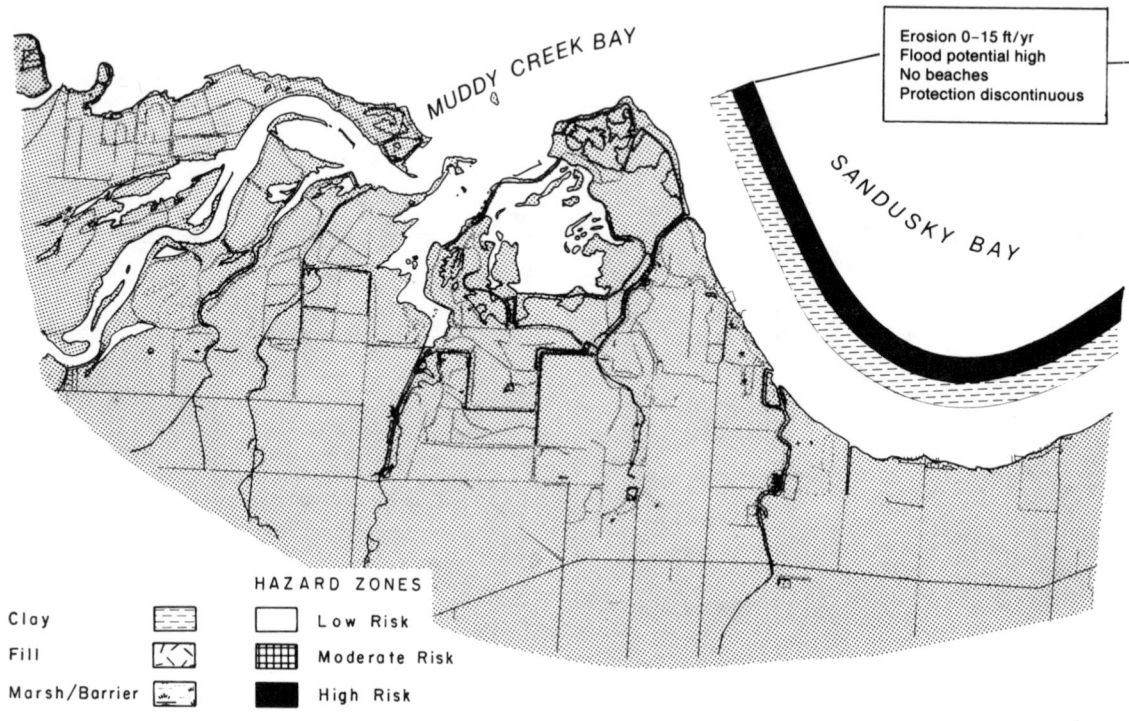

MUDDY CREEK BAY

SANDUSKY BAY

Erosion 0-15 ft/yr
Flood potential high
No beaches
Protection discontinuous

Clay

Fill

Marsh/Barrier

HAZARD ZONES

Low Risk

Moderate Risk

High Risk

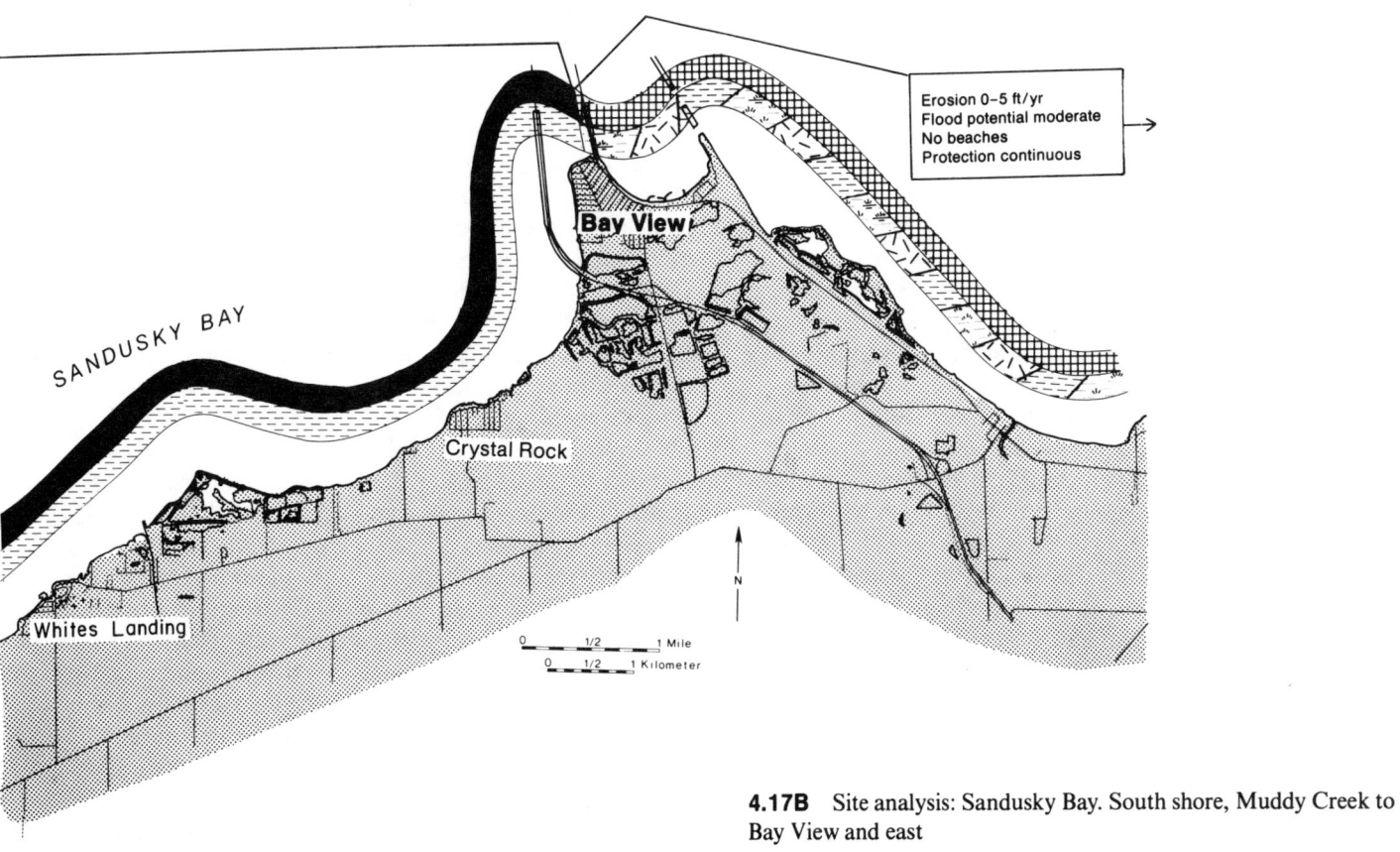

Erosion 0–5 ft/yr
Flood potential moderate
No beaches
Protection continuous

SANDUSKY BAY

Bay View

Crystal Rock

Whites Landing

0   1/2   1 Mile

0   1/2   1 Kilometer

N

4.17B   Site analysis: Sandusky Bay. South shore, Muddy Creek to Bay View and east

walls that are adequately maintained. Flooding is a major problem along this stretch. The flooding is often most severe in the Upper Bay. Homes constructed along this stretch should be floodproofed.

Erie County, Ohio

Stretch 18: Kelleys Island (fig. 4.16). This island, like the other rockbound Lake Erie islands, is an erosional remnant of the Columbus limestone. Weathering and erosion of the rock making up the island proceeded at slower rates than on the surrounding rock that has been eroded. The island is all that remains of the band of Columbus limestone that extended from Marblehead Peninsula to Pelee Island, Canada. The rock erodes slowly, but in fits and starts; undercutting is slow, collapse rapid. However, the till that caps the rock along much of the shore erodes much more rapidly. As a result, low-lying benches of limestone form along much of the eastern shore of the island where the till was eroded from the limestone. The island is known for its glacial grooves (elongate depressions related to glaciers) that traverse the island from east to west, and for its inscription rock, a slab of limestone along the south side of the island upon which hieroglyphics were carved by Native Americans.

Erosion rates are low, generally less than two feet per year. The rates are highest along the north and east sides of the island where the greatest wave energy reaches the shore or where till overlies limestone at lake level as at Kelleys Island State Park.

Flooding can be a problem, particularly along the north and east sides of the island where the relief is lowest and where the largest storm waves strike. The perimeter roads are flooded at times and low ridges of limestone gravel paralleling the shore attest to the infrequent but strong northeasters that cause marked storm surges and large waves.

Stretch 19: Cedar Point to Old Woman Creek (fig. 4.18). This stretch includes the point across from Bay Point (Cedar Point) as well as a short segment of the shore to the east of the point. Cedar Point is a bay-mouth spit (a sand spit that fronts a bay) that gradually built northwestward from the eastern margin of Sandusky Bay as longshore currents deposited sand from the net east-to-west sand transport system. The sand barrier is easily eroded and its low relief (generally 5 to 10 feet) makes it susceptible to flooding. The spit is largely built up from the chaussee entrance to the Cedar Point amusement park, which is visited by about 3 million people each year between late May and early September.

Prior to the record-high lake levels of the early 1970s the point was largely unprotected by man-made structures, but consisted of sand dunes covered by willows and cottonwoods. The intense storms in the early 1970s severely eroded the spit, resulting in the construction of extensive shore protection structures (fig. 4.19) particularly near the chaussee entrance and the causeway entrance. Although there was little structural damage to the homes along the point because they were built landward of the foredunes, flooding from the bay side occurred, as well as sand washover from the lake side. The narrowest segment of the spit at the east end has been cut through several times. The major breach, just east of the chaussee entrance, is likely to be permanent, as the ends of the spit on either side of the breach are now several hundred feet apart. If high lake levels persist and the supply of sand continues to decrease (the longshore transport of sand is effectively blocked by the Huron jetties) the spit will become an island connected to the mainland only by the chaussee and the causeway.

The segment of shore between Old Woman Creek and Cedar Point spit consists largely of 10-to-15-foot-high clay banks that are commonly fronted by man-made structures. Man-made structures run nearly continuously along this shore except for a half-mile-long segment east of the Huron River where the easternmost jetty has trapped a large quantity of sand, providing excellent shore protection. In general, erosion is not a great problem simply because of the man-made structures. But, even where there are structures, if homes are built too close to the lake (as along the shore side of the Huron River) there can be wave damage as well as foundation damage if the shore structures are poorly designed or built, or if maintenance has not been continued.

A major problem along the west Huron shore, and a related problem for the entire Cedar Point spit, is the scarcity of sand beaches to act as a buffer against storm waves. Unless the sand that has been transported to the west is replaced, even the beaches at Cedar Point will be gone with time. This is because the shore

protection structures are preventing erosion of the shore, and the Huron jetties block whatever sand is in the longshore transport system. Because of the importance of these beaches for shore protection and recreation we predict that there will be political pressure brought to bear on the Army Corps of Engineers for a beach nourishment program along this stretch of shore in the not-too-distant future. ˙

Erosion rates average about one foot per year along the Cedar Point spit except for the easternmost segment where they have been about ten feet per year. Between Rye Beach Road and Old Woman Creek, the rates average about two feet per year with the highest rates occurring along the shore just west (downdrift) of the jetties. Erosion has been thwarted by seawalls along much of the stretch. Especially effective are those constructed of heavy riprap, often with a concrete cap. The rough surfaces and a gentle (20 to 30 degree) slope help dissipate wave energy. Naturally these structures need periodic maintenance to make sure they are not being undercut or overtopped. Flooding can be a problem during extremely high lake levels from the Rye Beach Road area to the tip of Cedar Point because of the low relief. Homes should be floodproofed, especially if they are only a few feet above the long-term average lake level, because storm surges in this area can raise the lake three to four feet.

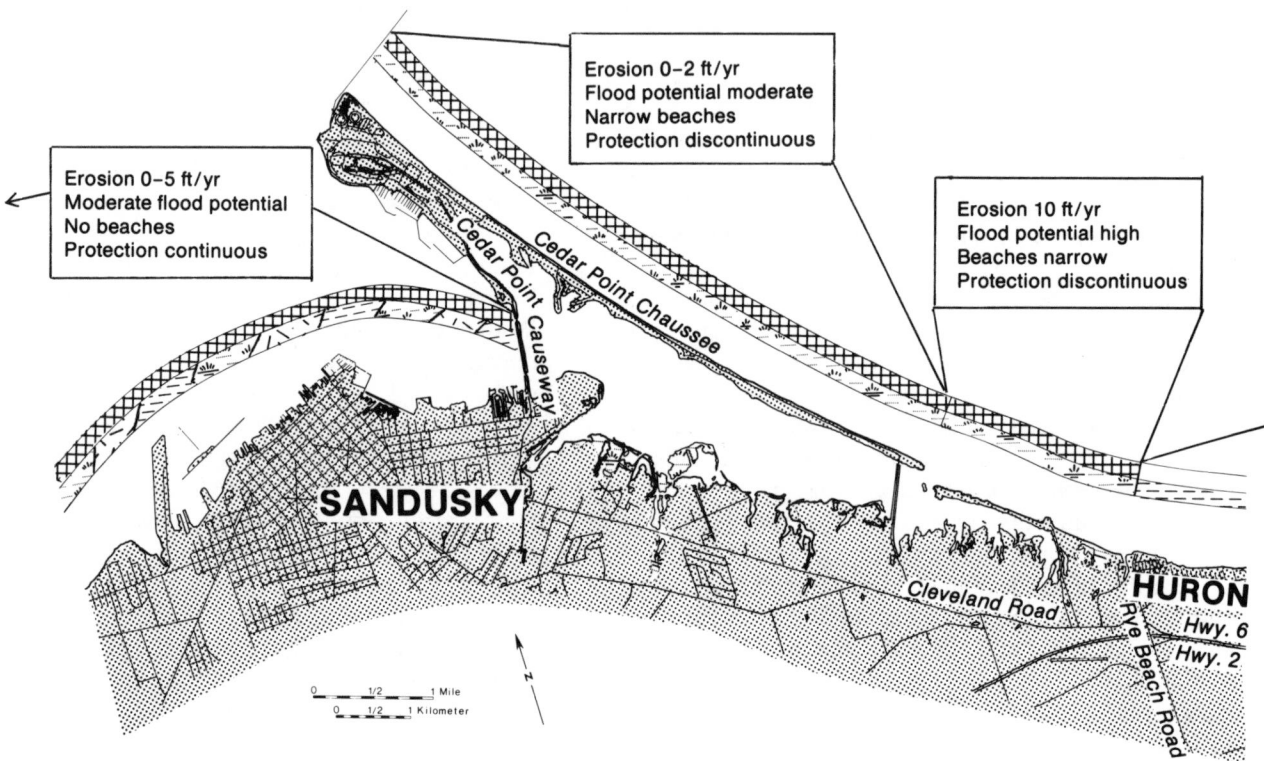

Erosion 0–5 ft/yr
Moderate flood potential
No beaches
Protection continuous

Erosion 0–2 ft/yr
Flood potential moderate
Narrow beaches
Protection discontinuous

Erosion 10 ft/yr
Flood potential high
Beaches narrow
Protection discontinuous

Cedar Point Causeway

Cedar Point Chaussee

SANDUSKY

HURON

Cleveland Road

Rye Beach Road

Hwy. 6

Hwy. 2

0   1/2   1 Mile
0   1/2   1 Kilometer

N

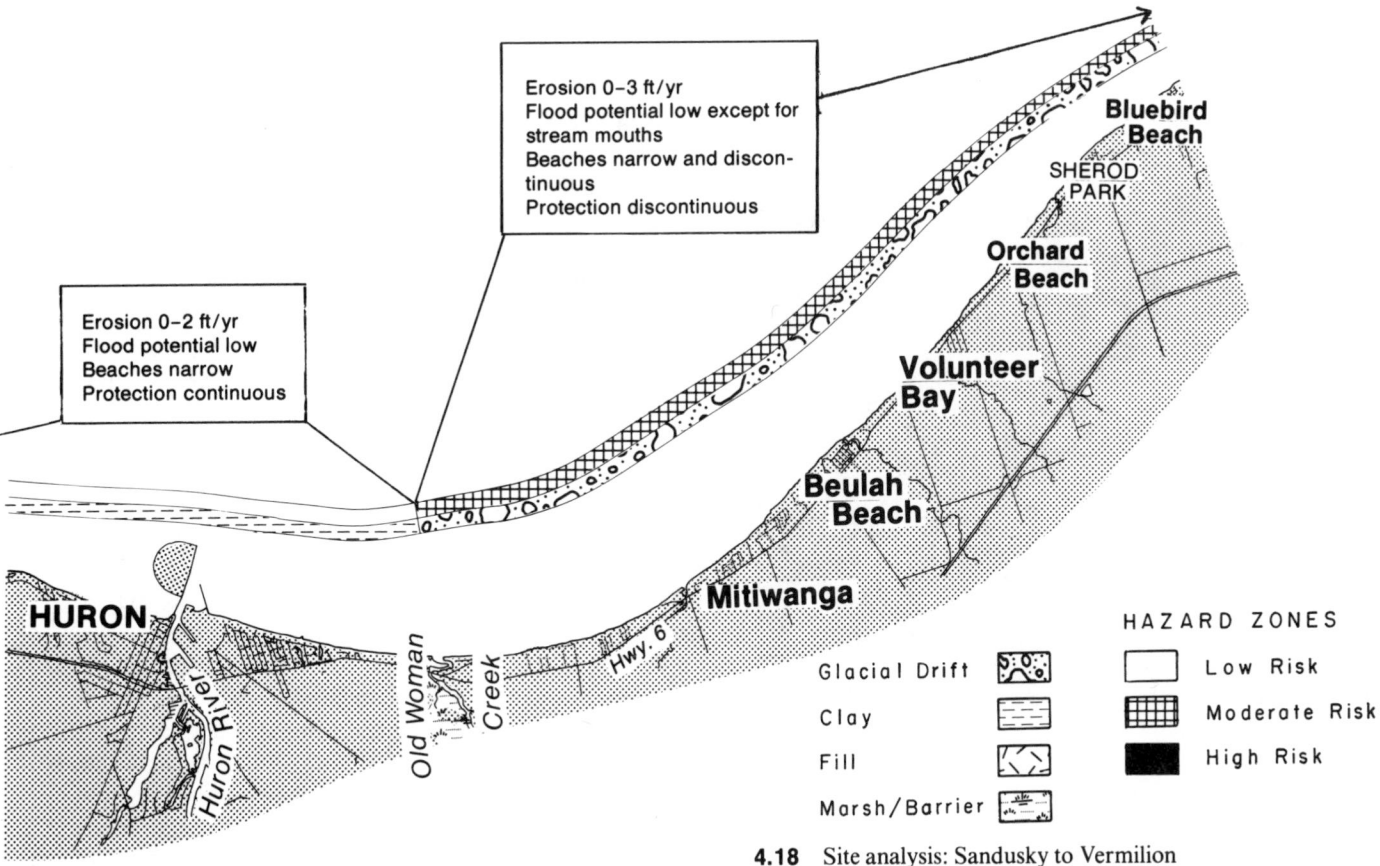

Erosion 0–3 ft/yr
Flood potential low except for
stream mouths
Beaches narrow and discon-
tinuous
Protection discontinuous

Erosion 0–2 ft/yr
Flood potential low
Beaches narrow
Protection continuous

Bluebird Beach

SHEROD PARK

Orchard Beach

Volunteer Bay

Beulah Beach

Mitiwanga

Hwy. 6

HURON

Huron River

Old Woman Creek

HAZARD ZONES

Glacial Drift	Low Risk	
Clay	Moderate Risk	
Fill	High Risk	
Marsh/Barrier		

4.18   Site analysis: Sandusky to Vermilion

4.19 Cedar Point. The Cedar Point barrier fronts East Bay (Sandusky). Relatively wide beaches and dunes help protect houses along the barrier, but narrowing beaches and high lake levels pose future flood and erosion problems.

Reach 2 (U.S.): Huron, Ohio (Old Woman Creek), to Erie, Pennsylvania (Presque Isle)

This reach, along the south central shore of Lake Erie (fig. 1.1), is characterized by moderate to high relief and clay or rock bluffs. The rock is Devonian shale that outcrops between Lorain and Cleveland, and the clay is till, a Pleistocene glacial deposit, that outcrops mainly between Huron and Lorain, and Cleveland and Erie. Because of greater relief and smaller storm surges, flooding is not the serious problem here that it is along the western reach. However, where people have built near or on a stream floodplain adjacent to the lake, as at Vermilion and Eastlake in Ohio, there can be major flood damage.

Although waves are bigger than at the west end of the lake, smaller storm surges and more resistant shore deposits contribute to lower erosion rates. In general, sand is scarce in the beach-nearshore system, particularly along the rockbound shore. However, in stretches along the till shore, particularly those between rocky headlands as between Huron and Vermilion, Avon Point and Lakewood, and even along exposed stretches as at Geneva-on-the-Lake, beaches are up to 100 feet wide in places.

The overall narrow and fragmented nature of the beaches, however, is a result of man-made structures. Jetties, as at Huron and Ashtabula, have essentially blocked the longshore transport of sand. This has caused the buildup of appreciable quantities of sand for a short distance on the updrift side of the structures, but has also caused erosion of the beaches for long distances on the downdrift side of the structures. These major disruptions to the longshore transport of sand have magnified the damaging effect of smaller, protective structures—as the beaches become narrower, the effect of a shore protection structure on adjacent property becomes greater. To try to prevent shore erosion, which is accelerated by the harbor jetties, shore protection structures are built. Groins in turn act as small jetties, trapping and altering the longshore transport of sand. Seawalls accelerate the longshore transport of sand and reduce erosion, which reduces the amount of sand coming into the system, causing the beaches to become even narrower.

Erosion rates are generally much lower here than along the western reach. The shale commonly erodes at a rate of less than 1 foot per year, and the till at 1 to 2 feet per year. Rates are higher if the shale is densely fractured or faulted, or if the till is unprotected by beaches. Stretches in the Grand River area have eroded at the rapid rate of ten feet per year once the protective beaches were eroded downdrift of harbor structures.

This reach is much more difficult to protect than the reach to the west because of the greater relief. When the bluffs and slopes are a few to several tens of feet high, greater earth forces are brought to bear on the shore protection structures. Not only does the toe of the slope need to be stabilized, the stability of the entire slope needs to be considered. For the clay bluffs and slopes, a substantial seawall at the toe is generally required, and a gentle or terraced slope behind the wall is necessary for even temporary slope stabilization. In zones that transmit water through the clay,

drains should be installed to intercept the water before it reaches the free face of the slope. This type of problem is particularly likely in zones of sand and/or sand and clay between tills or on top of a till. In such areas, weep holes may be needed in the seawall to allow water to escape and to prevent excessive fluid pressure from causing wall failure. Plants should cover the surface to help hold the soil and prevent erosion by runoff at the surface. There are many seawall designs. The trick is to find a type that can withstand both wave impact and erosion as well as slope forces. A good seawall should have a surface that will dissipate wave energy so that beach sand will not be scoured from the toe by wave action. We are generally not advocates of groins, but if there is a reasonable sand supply in the beach-nearshore system, short (20 to 30 feet), low groins will impede the longshore transport of sand, which helps protect the toe of the slope, and provides a beach.

"Providing a beach," however, should be put into context. If buildings had not been constructed too close to bluff edges and seawalls constructed to protect such structures, natural beaches would still be found at the base of the bluffs.

Erie and Lorain Counties, Ohio

Stretch 1: Old Woman Creek to Lorain (figs. 4.18 and 4.20). This stretch is characterized by 20-to-30-foot-high bluffs of till or shale bordered by narrow, discontinuous beaches. The shore is mostly developed, although there is still some open land. A large

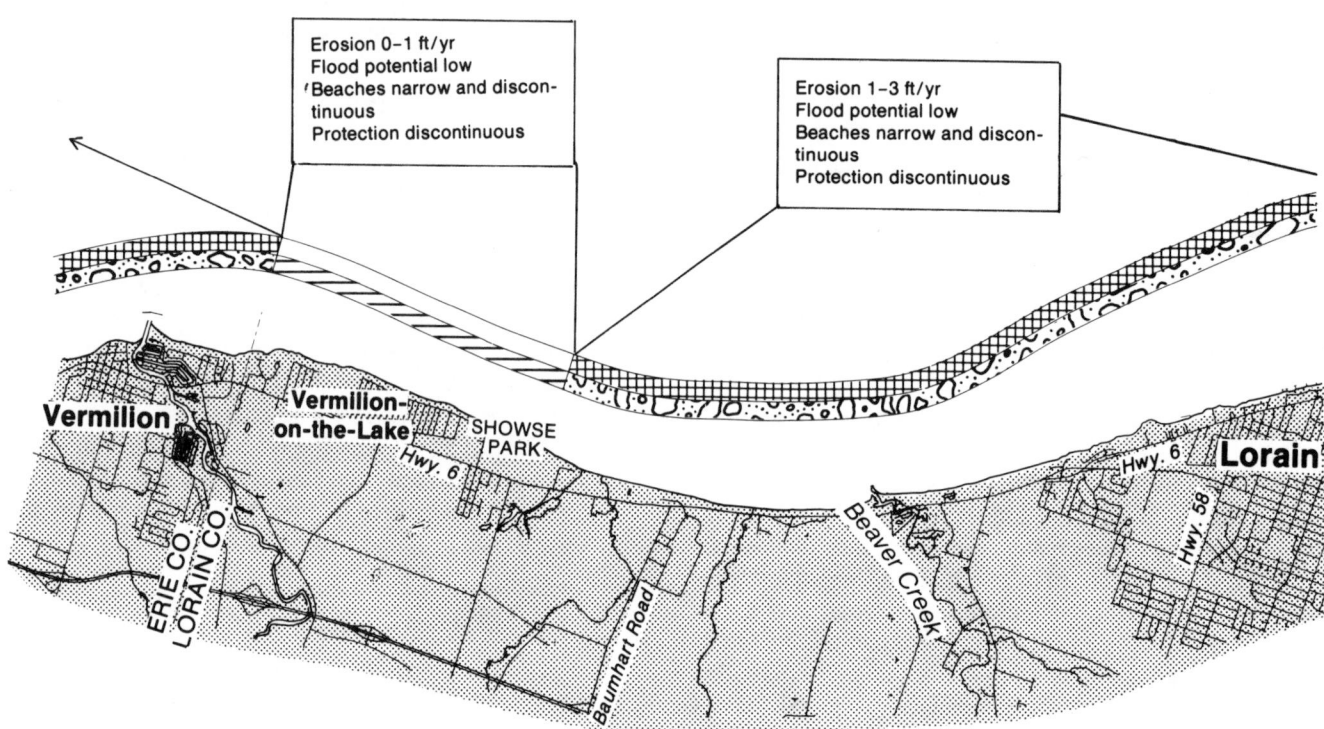

Erosion 0–1 ft/yr
Flood potential low
Beaches narrow and discontinuous
Protection discontinuous

Erosion 1–3 ft/yr
Flood potential low
Beaches narrow and discontinuous
Protection discontinuous

Vermilion

Vermilion-on-the-Lake

SHOWSE PARK

Hwy. 6

Lorain

Hwy. 6

Hwy. 58

ERIE CO.
LORAIN CO.

Baumhart Road

Beaver Creek

4.20   Site analysis: Vermilion to Avon Point

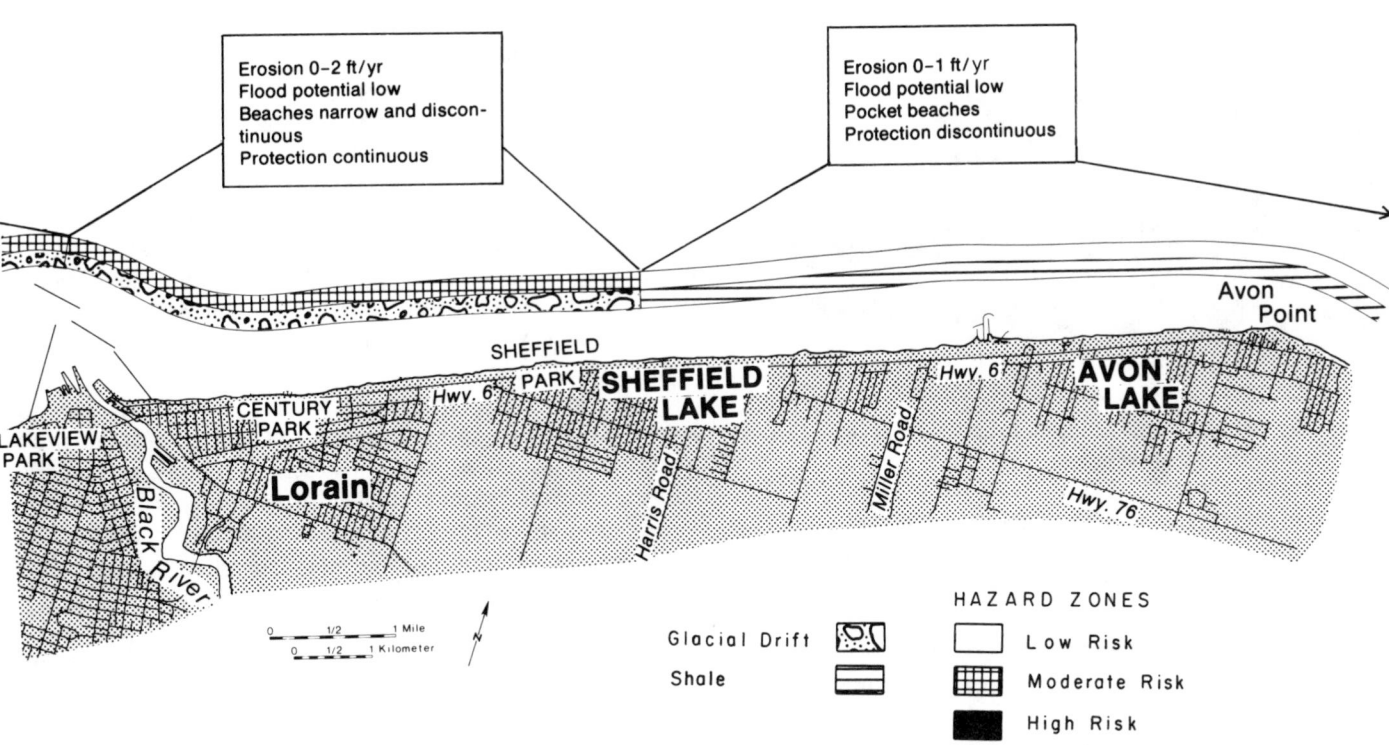

Erosion 0–2 ft/yr
Flood potential low
Beaches narrow and discontinuous
Protection continuous

Erosion 0–1 ft/yr
Flood potential low
Pocket beaches
Protection discontinuous

Avon Point

SHEFFIELD
PARK
CENTURY
PARK
Hwy. 6
SHEFFIELD
LAKE
Hwy. 6
AVON
LAKE
LAKEVIEW
PARK
Lorain
Black River
Harris Road
Miller Road
Hwy. 76

0   1/2   1 Mile
0   1/2   1 Kilometer
N

Glacial Drift
Shale

HAZARD ZONES
Low Risk
Moderate Risk
High Risk

4.21 Shore protection, Lorain County. Note the variety of types. Construction continues along the lakeshore.

number of shore protection structures have been built along this stretch (fig. 4.21). Erosion is the principal hazard as the shore is above storm surge level except for the floodplains of the small creeks that enter the lake.

The shore was developed in the early 1900s by people from Cleveland and the northern Ohio area who were looking for places to spend their summer vacations. Place names such as Bluebird Beach and Orchard Beach reflect this early use, but in the past 10 to 20 years, the shore has become increasingly populated by permanent residents. Along with development came

shore protection structures as the sparse sand supply and associated narrow beaches afforded little natural protection. The groins and seawalls modified the natural movement of sand, and the continuous beaches that were present before the structures were built quickly became discontinuous. Naturally, the larger the structure the greater its effect, and this is particularly true of the nearly 1,000-foot-long Vermilion jetties that were built in the mid-1830s. Not only did the jetties block the flow of sand, robbing the beaches to the west, but they diverted sand offshore, probably far enough so that the sand was lost to the longshore transport system. This decrease in sand supply partly explains the riddle of why there are communities with beach names, but without beaches.

Erosion rates range from 0 to 3 feet per year. The rates are lowest along the rockbound shore east of Vermilion and highest along the till shore both east and west of Vermilion, particularly along unprotected segments downdrift of groins. In general, the more closely spaced the shore protection structures, the lower the erosion rates. Because of the moderate relief, flooding is not a problem along most of the lakeshore, but there are problems where people have built on the stream floodplains adjacent to the lake. In particular, homes built on the Vermilion floodplain were flooded in the early 1970s because of high lake levels and associated northeast wind storms.

Shore erosion protection, because of the greater relief, is more of a problem along this stretch (and the shore to the east) than along the low-relief shore to the west. This difference is primarily the result of the greater downslope force exerted by the higher (more massive) slopes and bluffs. The crucial area to consider in attempting to stabilize such slopes is the toe of the slope or bluff. If the toe is unprotected or unsupported, downslope movement (mass wasting) will continue as long as the lake erodes the material at the toe. Because of the sparse supply of sand along this stretch, groins should not be built, particularly if the down-drift (west) shore is unprotected. The best form of protection is a gentle, landscaped slope, fronted at the toe by heavy riprap. The riprap will protect the toe of the slope from erosion, and the gentle slope will reduce the downward (driving) force acting on the riprap. The beach will eventually be sacrificed in protecting the property. Vertical seawalls can reduce erosion rates for a while, but they also reflect wave energy and prevent the formation of a beach. Moreover, walls can easily fail if they are not strong enough to resist the slope forces acting on them. Naturally, any structure will have to be maintained, and anyone locating in this area should consider the long-term costs of permanent maintenance of the stabilization structures.

Lorain and Cuyahoga Counties, Ohio

Stretch 2: Lorain to Bradley Road (figs. 4.20 and 4.22). This stretch, like the Old Woman Creek to Lorain stretch, is characterized by 20-to-40-foot-high bluffs of till or shale, though unlike the area to the west, shale predominates. Narrow, discontinuous beaches front the till, whereas beaches are absent along the shale

shore except as small pocket beaches or adjacent to groins. Erosion is a major problem, especially in the till. The shale erodes too, but at a slower rate. Keep in mind, though, that a given erosion rate is only an average—which can be deceptive. Even where rates are slow, erosion commonly takes place as infrequent, but earth-shaking falls. The process is a simple one: the shale is slowly undercut by waves at lake level until the overlying rock lacks the strength to support itself and collapses. Cracks (joints) reduce the strength of the shale. Most of the man-made shore protection structures have been built along the predominately till shore between Lorain and Sheffield Lake where groins trap sand in narrow beaches and numerous seawalls armor the shore.

East of Sheffield Lake to Bradley Road just inside Cuyahoga County, the shale shore is largely unprotected, except where there are homes quite close to the lake. The irregular shape of the shoreline along the shale bluffs east of Avon Point is due to small structural features in the rock called folds and faults. The shale has been intensely fractured by these folds and faults, and is thus more easily eroded by waves. This differential erosion produces the irregular headlands and embayments.

Erosion rates range from 0 to 2 feet per year. The rates are highest along the till shore and lowest along the shale shore. Again, a gentle, landscaped slope and adequate toe protection is probably the best form of shore protection for the till. For the shale, structures to protect the base of the bluff from wave erosion will be necessary if buildings are built too close to the edge. Breakwaters might be the best solution here. Groins are not

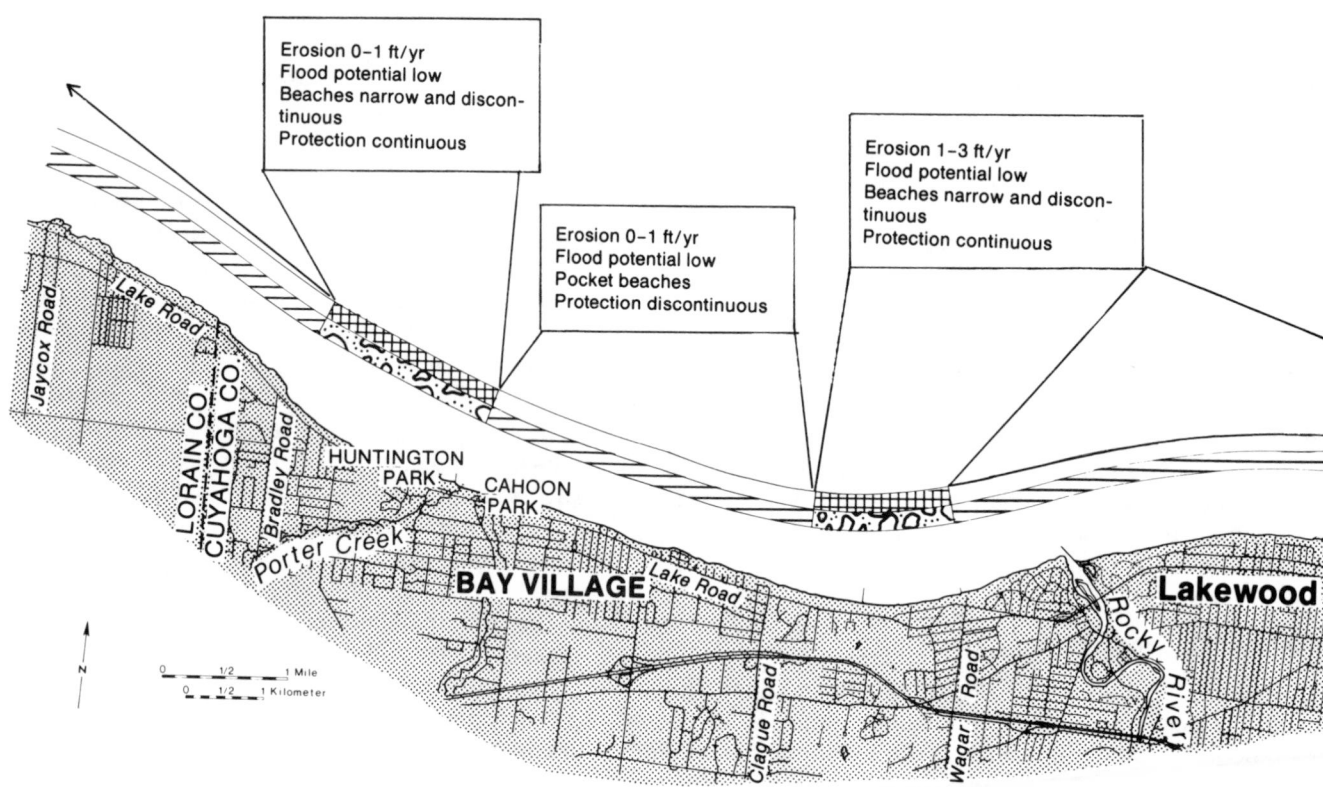

Erosion 0–1 ft/yr
Flood potential low
Beaches narrow and discon-
tinuous
Protection continuous

Erosion 1–3 ft/yr
Flood potential low
Beaches narrow and discon-
tinuous
Protection continuous

Erosion 0–1 ft/yr
Flood potential low
Pocket beaches
Protection discontinuous

4.22 Site analysis: Avon Point to Bratenahl

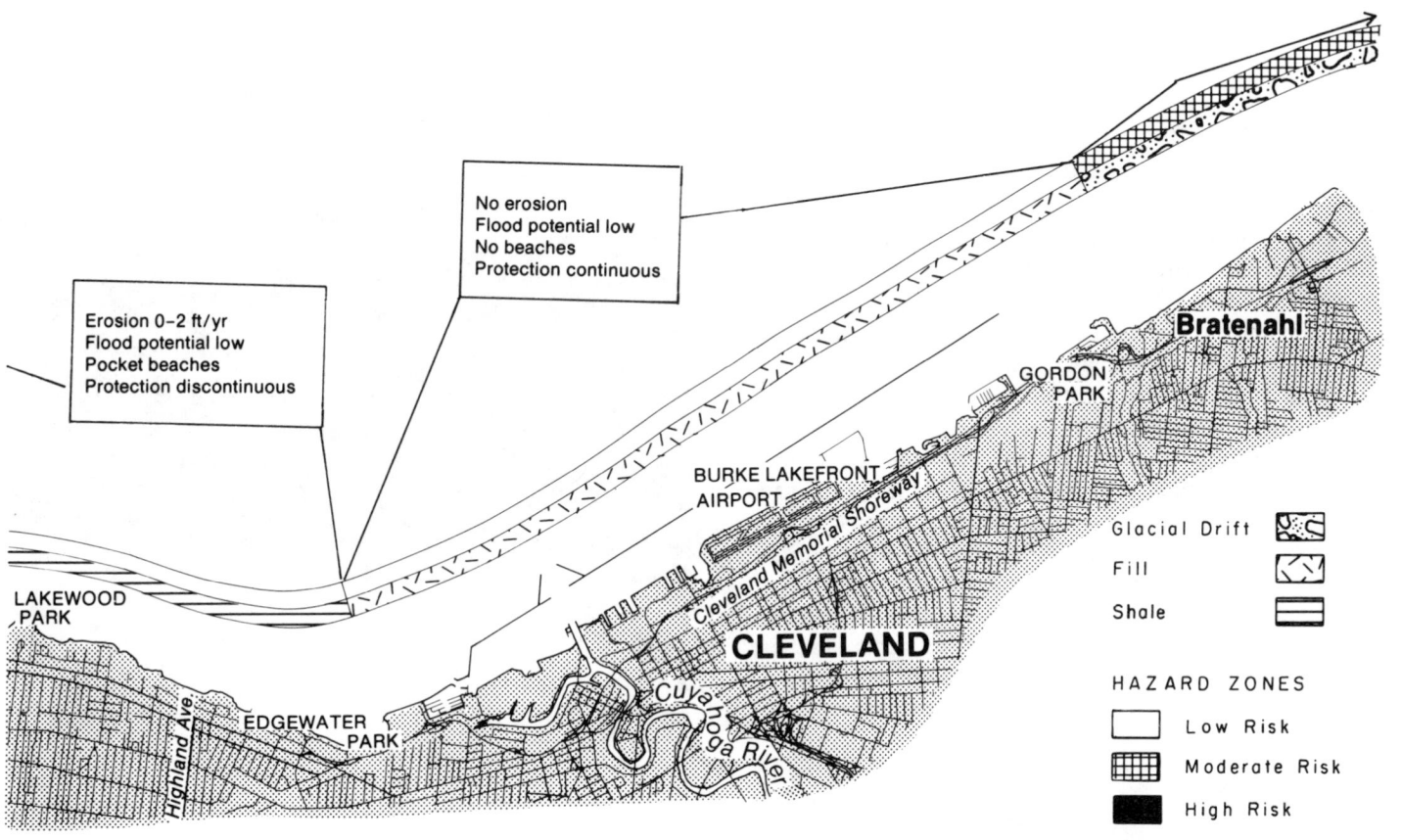

No erosion
Flood potential low
No beaches
Protection continuous

Erosion 0–2 ft/yr
Flood potential low
Pocket beaches
Protection discontinuous

Bratenahl

GORDON
PARK

BURKE LAKEFRONT
AIRPORT

Cleveland Memorial Shoreway

LAKEWOOD
PARK

CLEVELAND

EDGEWATER
PARK

Highland Ave.

Cuyahoga River

Glacial Drift

Fill

Shale

HAZARD ZONES

Low Risk

Moderate Risk

High Risk

recommended because of the small sand supply and adverse downdrift effect.

Cuyahoga County, Ohio

Stretch 3: Bradley Road to Edgewater Park (fig. 4.22). This stretch is characterized by 40-to-60-foot bluffs of shale or till. Short segments of fill land are present, such as at Lakewood Park and Edgewater Park. Fronting the till bluffs, which make up no more than 20 percent of the stretch, are narrow, discontinuous beaches. The shale bluffs have pocket beaches in the embayed areas. Man-made structures, mostly seawalls and groins, are concentrated along the more easily erodible till shore. Erosion problems were significant in the 1970s for segments made up of till, and signs reading "Fill Dirt Wanted" were common along Lake Road. Even along Huntington Park, whose long (up to 200 feet) groins have trapped 50-foot-wide beaches, minor slumping took place as storm waves eroded till from the toe of the slope, and the bluffs slumped because of decreased resistance at the toe. The stretch is nearly completely developed. Major problem areas include the two moderate-risk till bluff segments and those shale bluffs where structures are built too close to the lake (fig. 4.23). In either situation the base of the bluff needs to be protected from wave erosion.

Erosion rates range from 0 to 2 feet per year except for the short segment of till between Clague and Wagar Roads where the rates are closer to 3 feet per year. Overall, the shore protec-

4.23 Rock fall at Bay Village. Fractures in rock are exploited by wave forces that quarry the fragments. Loss of support leads to bluff failure even in rock.

tion structures and beaches in the Huntington Park area have kept this till shore from receding more rapidly than the shale-fronted shore. The high relief along this stretch requires that such structures be massive enough to prevent downslope movement, and they are therefore expensive to build and maintain. When a building has been built too close to the lake on the shale, the base of the shale must be protected so that it will not be undercut and collapse. The long-term cost of such protection may be higher than the cost to move the building back out of harm's way. Because of the high relief, flooding has not been a problem in this area.

Stretch 4: Cleveland lakefront (fig. 4.22). This area is fill land. The fill extends out into the lake from the bluffs south of the Cleveland Memorial Shoreway. Material for the fill, which began to be placed in the 1800s and is continuing to be placed in dike disposal areas today, has come from trash and excavations in the city and from sediment dredged from the mouth of the Cuyahoga River. The stretch is armored with seawalls and is fronted for the most part by a nearly six-mile-long dimension stone breakwater. There are no beaches except at Edgewater Park at the west end of the lakefront. Cleveland has reached the ultimate urbanized shoreline stage.

Cuyahoga and Lake Counties, Ohio

Stretch 5: Bratenahl to Chagrin River (fig. 4.24). Forty-foot-high till bluffs (in places capped by sand), separated by twenty-

foot bluffs of till and shale make up this shore. Fronting the bluffs are narrow, discontinuous beaches (fig. 4.25). Most of the beaches are located in shallow embayments that have formed along unprotected stretches of shore. Moreover, the beaches at White City Park provide the first indication of the net west-to-east longshore transport of sand east of Cleveland. At the park a wide (up to about 400 feet) beach extends along the west side of the structures and a narrow (less than 50 feet) beach occurs along the east side of the structures. Man-made structures, which consist mainly of seawalls and groins, front the till shore. Former seawalls, now detached from the shore because of erosion behind the walls, act as partial breakwaters in places. Some of the longest, most continuous concrete seawalls to be found along the lakeshore are located along this stretch. These seawalls reflect wave energy, which scours sand from their bases, preventing the buildup of beach sand and thus the formation of beaches. The most significant erosion problems occur along the eastern one-third of shore where groins have led to accelerated erosion along unprotected segments of the downdrift shore. Many lakeshore homes are so close to the lake that owners are unable to landscape the slopes in front of them, and major structural measures are necessary to preserve the homes, even temporarily. The best solution, if possible, is probably to move the houses back away from the retreating slope edge. A small area in Eastlake, built on the Chagrin River floodplain, suffered major damage in the storms of 1972 and 1973 (fig. 4.26). The U.S. Army Corps of Engineers, at a cost of about $1.5 million, constructed dikes in

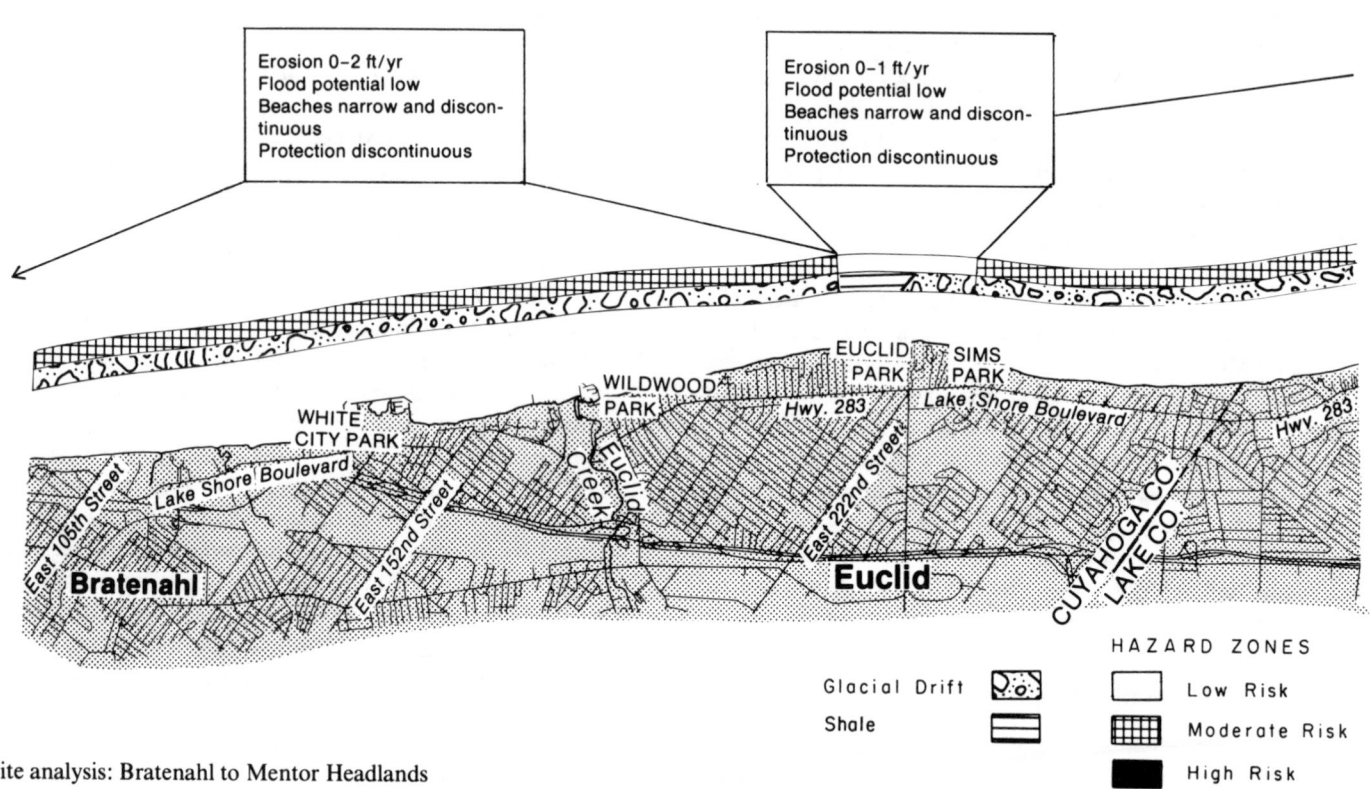

Erosion 0–2 ft/yr
Flood potential low
Beaches narrow and discontinuous
Protection discontinuous

Erosion 0–1 ft/yr
Flood potential low
Beaches narrow and discontinuous
Protection discontinuous

HAZARD ZONES

Glacial Drift

Shale

Low Risk

Moderate Risk

High Risk

4.24 Site analysis: Bratenahl to Mentor Headlands

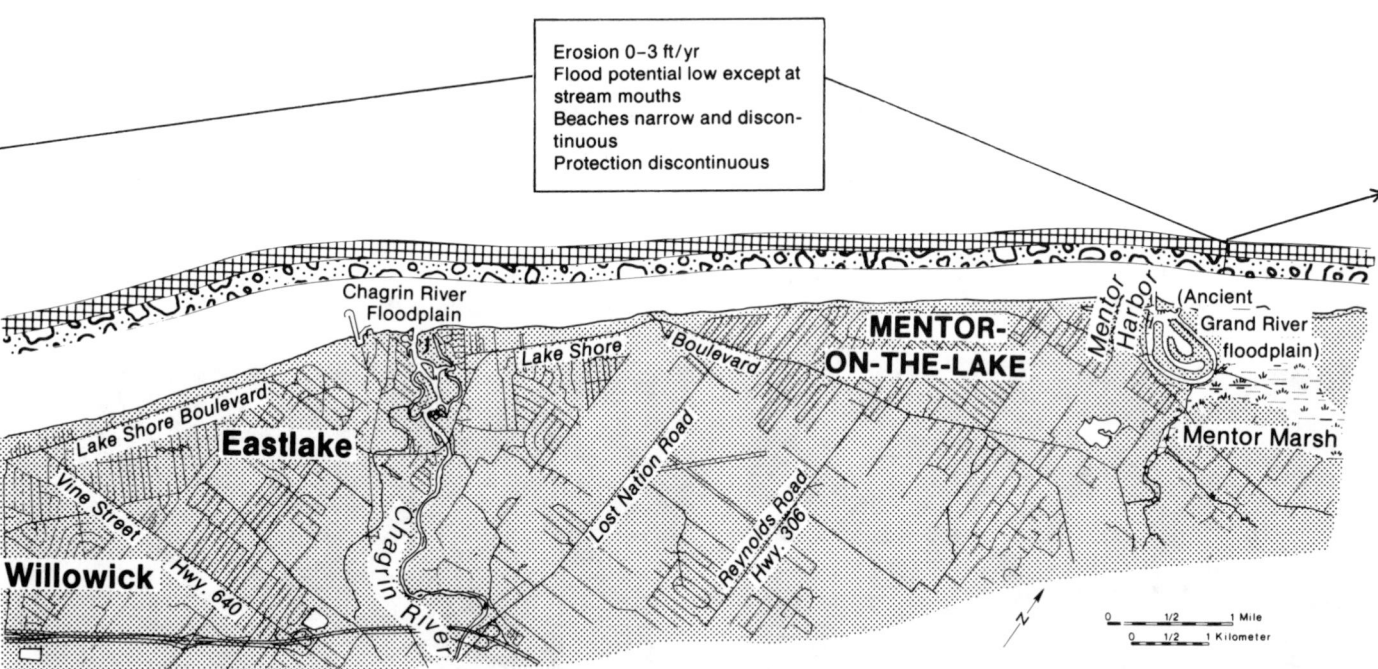

Erosion 0–3 ft/yr
Flood potential low except at stream mouths
Beaches narrow and discontinuous
Protection discontinuous

Chagrin River Floodplain

Lake Shore Boulevard

MENTOR-ON-THE-LAKE

Mentor Harbor

(Ancient Grand River floodplain)

Mentor Marsh

Lake Shore Boulevard

Eastlake

Lost Nation Road

Reynolds Road Hwy. 306

Vine Street

Willowick   Hwy. 640

Chagrin River

N

0   1/2   1 Mile
0   1/2   1 Kilometer

the 1970s to prevent flooding of this area. The estimated damage prevented was about $1.7 million. The majority of the homes in the floodplain are still occupied, although there are no longer any homes adjacent to the dike.

Erosion rates range from 0 to 3 feet per year with the highest rates along the broad shallow embayment east of the shale headland at Euclid Park. The erosion rates may be higher here because the shore has not been as heavily armored as the shore fronting the Bratenahl area. The key to long-term protection is a stable toe (base of slope) that will withstand both wave forces on the lake side and downslope (earth) forces from the shore side. The homes in the Chagrin River-Lake Erie floodplain should be floodproofed because this is an open, exposed location, one likely to flood.

4.25 Shore erosion in the Euclid area. Wave erosion at the toe of the slope leads to an unstable slope and slope failure. Note the narrow beach characteristic of this area.

4.26 Storm damage at Eastlake. Double trouble. Homes here are subject to flood damage from the Chagrin River as well as to flood and erosion damage from Lake Erie.

Lake County, Ohio

Stretch 6: Chagrin River to County Line Road (figs. 4.24 and 4.27). This moderate risk shore is one of the most diverse along Lake Erie. The till bluffs, which are capped by silt and sand in places, rise in relief from 20 to 30 feet in the Chagrin River area to 40 to 50 feet in the Grand River area and east to Redbird, where the relief drops to only 10 to 20 feet above the lake. The shore lacks the rock that outcrops to the west, and in the till, particularly in the Grand River area, are laminated silts and clays that contribute to greater slope instability by acting as slip planes within the till.

The beaches are diverse. For the most part they are narrow and discontinuous, but the jetties at Mentor Harbor and Fairport Harbor have trapped sand and built beaches on their updrift sides. The Fairport Harbor jetties have caused the shoreline to build out about 2,000 ft into the lake on the west side of the structures since the mid-1820s. Unfortunately, such structures have caused an uneven distribution of sand along the shore, helping to protect some shore segments, but leaving greater lengths of shore unprotected, thus leading to accelerated erosion. Erosion rates are as high as 6 to 7 feet per year along the downdrift (east) segments of Mentor Harbor and Fairport Harbor. This stretch includes the first long segment (Painesville-on-the-Lake to Redbird) of largely undeveloped shore east of Cleveland. However, there are a few large industries such as the conspicuous Perry nuclear plant. Shore protection structures are less common here, although they do front significant parts of the shore between the Chagrin River and Grand River, and along Redbird and Madison-on-the-Lake.

Erosion rates are generally 0 to 3 feet per year, but they may be as high as 6 to 7 feet per year. Major erosion hazards exist at Mentor Headlands and Painesville-on-the-Lake (fig. 4.28). Along Mentor Headlands, the narrow beaches backed by till interlayered with silts and clays make the shore slopes susceptible to failure. The unstable setting has resulted in the loss of homes and portions of Headlands Road. Along Painesville-on-the-Lake the loss of sand due to the Fairport Harbor jetties and groins at Township Park has led to the loss of beaches and to greatly accelerated erosion. This phenomenon is common along this stretch and ranges from major problem areas associated with jetties to minor ones associated with groins. Shore erosion, on the other hand, has contributed to the development of Fairport Harbor. Probably no more than a few hundred years ago the retreating lakeshore intersected a meander bend of the Grand River, which caused the Grand River to flow into the lake at Fairport Harbor rather than at Mentor Harbor. Mentor marsh is the former floodplain of the Grand River.

Ashtabula County, Ohio

Stretch 7: County Line Road to Conneaut (figs. 4.29 and 4.30). Till banks of about ten foot relief at the Lake-Ashtabula county line rise abruptly to 30-foot bluffs at Geneva-on-the-Lake

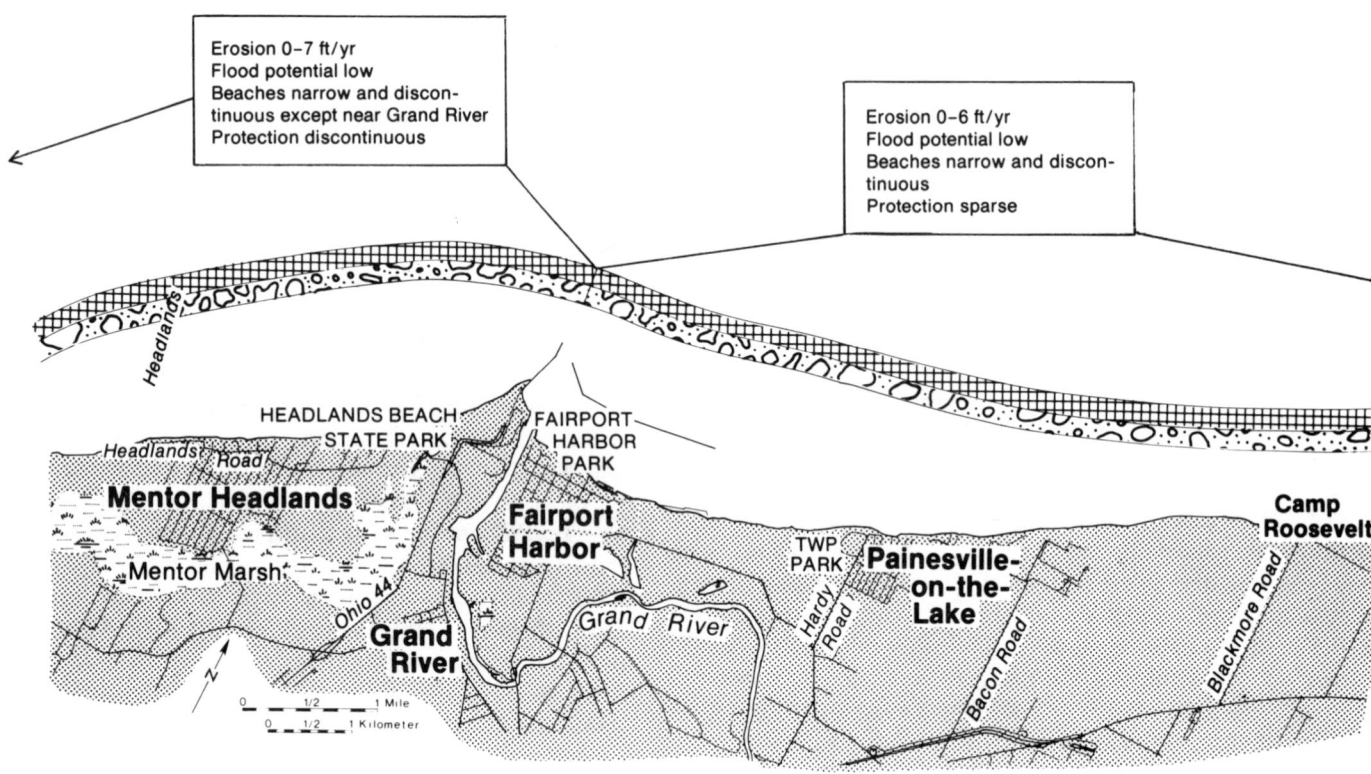

Erosion 0–7 ft/yr
Flood potential low
Beaches narrow and discon-
tinuous except near Grand River
Protection discontinuous

Erosion 0–6 ft/yr
Flood potential low
Beaches narrow and discon-
tinuous
Protection sparse

4.27 Site analysis: Mentor Headlands to the Lake-Ashtabula County line

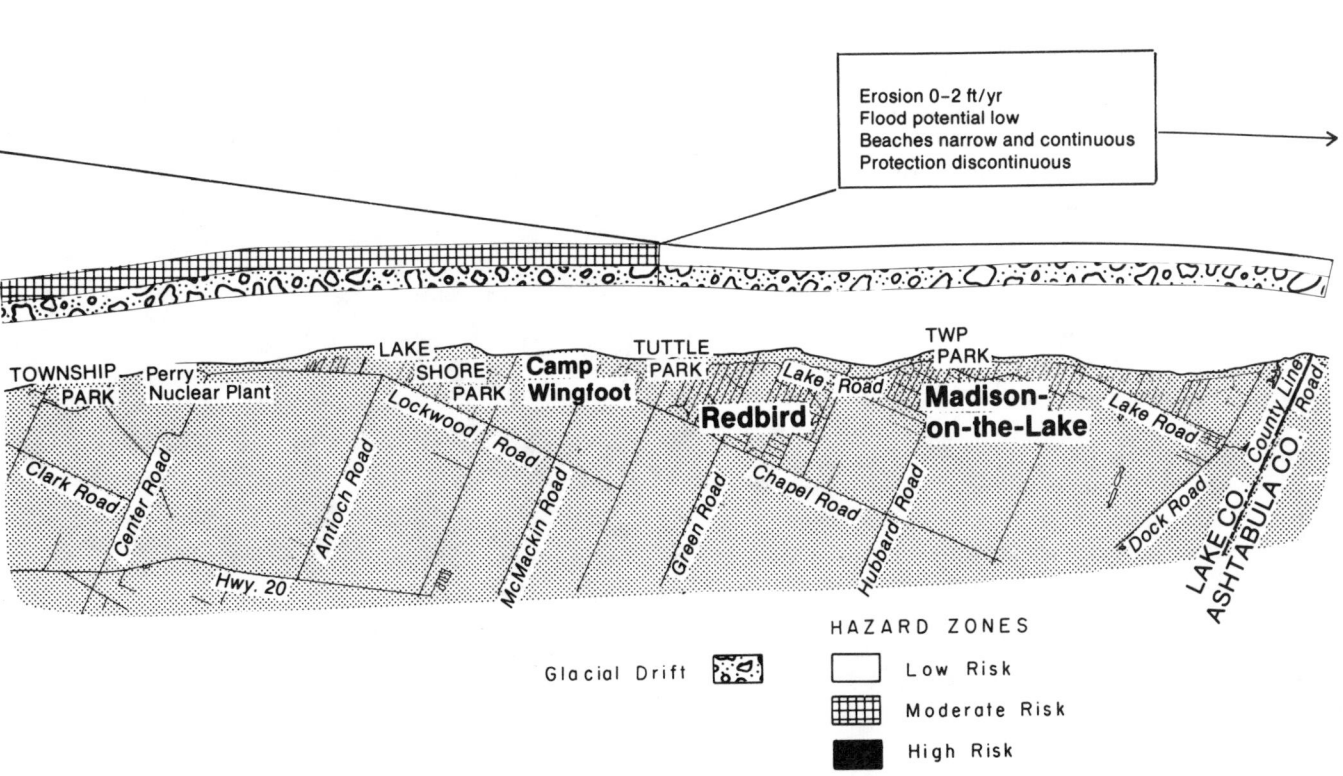

Erosion 0–2 ft/yr
Flood potential low
Beaches narrow and continuous
Protection discontinuous

TOWNSHIP PARK
Perry Nuclear Plant
Clark Road
Center Road
Hwy. 20
Antioch Road
Lockwood Road
McMackin Road
LAKE SHORE PARK
Camp Wingfoot
Green Road
TUTTLE PARK
Redbird
Chapel Road
Lake Road
Hubbard Road
TWP PARK
Madison-on-the-Lake
Lake Road
Dock Road
LAKE CO.
ASHTABULA CO.
County Line Road

Glacial Drift

HAZARD ZONES
Low Risk
Moderate Risk
High Risk

and continue to rise gradually to 70 feet just west of Conneaut. The till includes some interlayered silt and clay and is capped by silt and sand east of Ashtabula. Beaches are widest in the segment between the county line and Geneva-on-the-Lake, but rapidly become narrower to the east where they are interrupted by man-made structures. Large quantities of sand have been trapped to the west of the major harbor structures at Ashtabula and Conneaut as well as on the west side of a multitude of smaller jetties and groins that front the lakeshore. Erosion rates are usually higher along an unprotected shore downdrift of a jetty or groin. Shore protection structures are fairly common between the Lake County line and Ashtabula, and from Grant Street to Conneaut, but in the intervening segment there are only occasional isolated groins or seawalls.

Erosion rates range from 0 to 3 feet per year. The higher rates usually occur downdrift (to the east) of the jetties and groins. Overall, the average rate for the stretch is about 1 foot per year. Major slope stabilization was undertaken along Ohio Route 531 just east of Geneva-on-the-Lake (fig. 4.31) when the road was nearly captured by the lake. Future problems of a similar nature are likely along the same segment of shore as well as just west of Kingsville-on-the-Lake because the road was constructed too close to the lake. Small (less than 50 feet long) groins have helped reduce erosion along this stretch because there is a reasonably sufficient sand supply. However, if the groins or jetties are too large, such as the harbor structures and groins west of Ashtabula, then the downdrift beach does not receive enough sand and

4.28   Erosion and shore protection at Painesville-on-the-Lake. The debris placed on this slope is a futile attempt to forestall erosion ultimately caused by the Fairport Harbor jetties.

accelerated erosion takes place. Naturally, grading and draining the slope reduces the likelihood of major slumps and slides.

Ashtabula County, Ohio, and Erie County, Pennsylvania

Stretch 8: Conneaut to Presque Isle (figs. 4.30 and 4.32). Till bluffs of 40-foot relief at Conneaut rise gradually to 60 feet at Crooked Creek, and then climb rapidly to 80 feet east of Crooked Creek to about 100 feet near Elk Creek. From there, 80-to-90-foot bluffs persist to Erie. The till contains some interlayered silt and clay and is overlain by an appreciable thickness of silt and sand along much of the stretch. Shale outcrops near lake level in places about one to three miles east of Walnut Creek (Manchester Beach).

Narrow beaches front nearly the entire stretch. The beaches are widest near the largest streams such as Crooked Creek, Elk Creek, and Walnut Creek, in the more embayed segments of the shore. The streams transport sediment to the lake (unlike the Michigan and Ohio stream mouths, the stream mouths in Pennsylvania and New York are not drowned and thus transport much more sand to the lakeshore).

The streams also impede the longshore transport of sand, similar to jetties and groins, but in a much more subdued fashion. The scattered groins and jetties have trapped sand on their updrift (west) sides and for the most part have made wider beaches (fig. 4.33). Longer groins make up groin fields such as at Crooked Creek (Camp Fitch) and Trout Run where the sand is more abundant because of the sand transport to the lake by the streams. Overall, the groins have not had a major impact. Some have trapped only a small amount of sand. Others have trapped large amounts of sand, but have been constructed where sand is so abundant (such as east of Elk Creek or east of Walnut Creek) that the shore downdrift of these structures has not been markedly affected. In general, the shore is sparsely settled with summer camps, lodges, and homes.

Erosion rates range from 0 to 3 feet per year. For the most part, the rates are highest downdrift of the largest jetties and groins. Particularly along the western border of the stretch, the Conneaut Harbor jetties have had a significant influence in reducing the longshore transport of sand to Pennsylvania. Overall, erosion rates are lowest where the beaches are widest; however, the marked relief and the easily erodible shore deposits make erosion control along the shore a particularly trying endeavor. For this reason, many have built homes and cottages near the lower-relief stream mouths (where they may now be in danger of flooding) or a good distance away from the shore so that erosion will not have an effect on their homes for a few hundred years. Short groins are a partial solution as beaches are the best natural form of shore protection, but where beaches are relatively narrow, the waves are able to erode the base of the bluff, creating major problems with slope stability. The extreme height of the bluffs increases the problem. It is much easier and much less expensive to design a seawall for a 5-to-10-foot bank along the

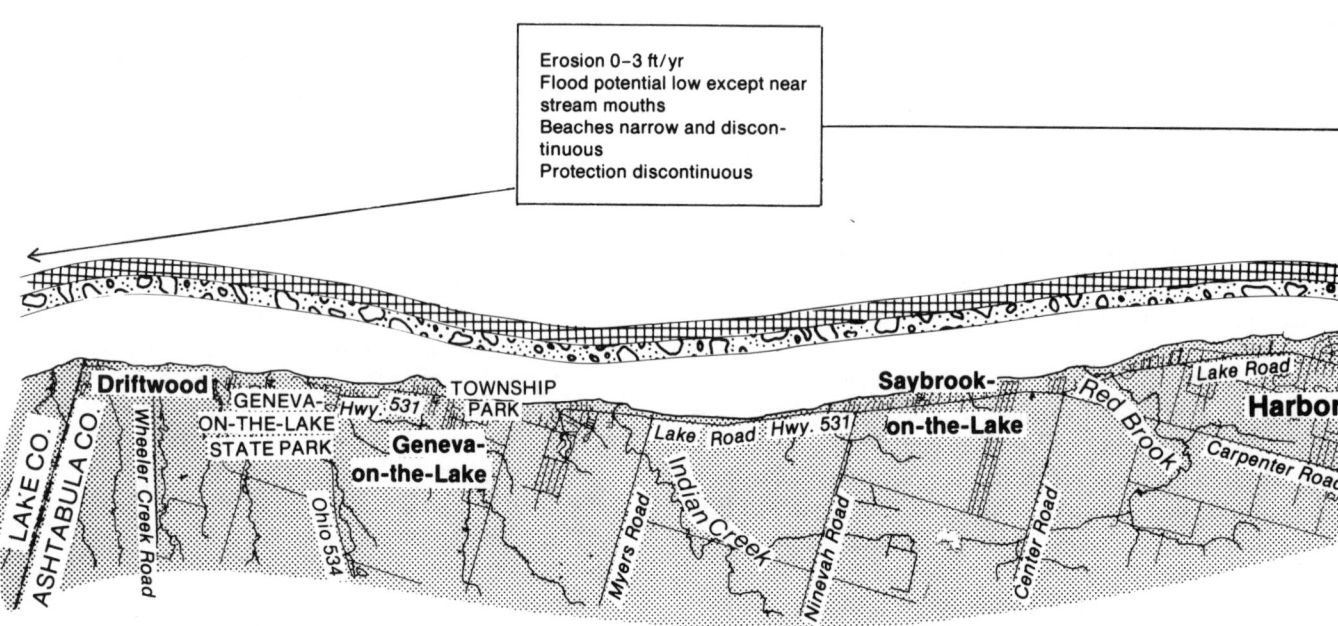

Erosion 0–3 ft/yr
Flood potential low except near
stream mouths
Beaches narrow and discon-
tinuous
Protection discontinuous

4.29   Site analysis: Lake-Ashtabula County line to Camp Calvary

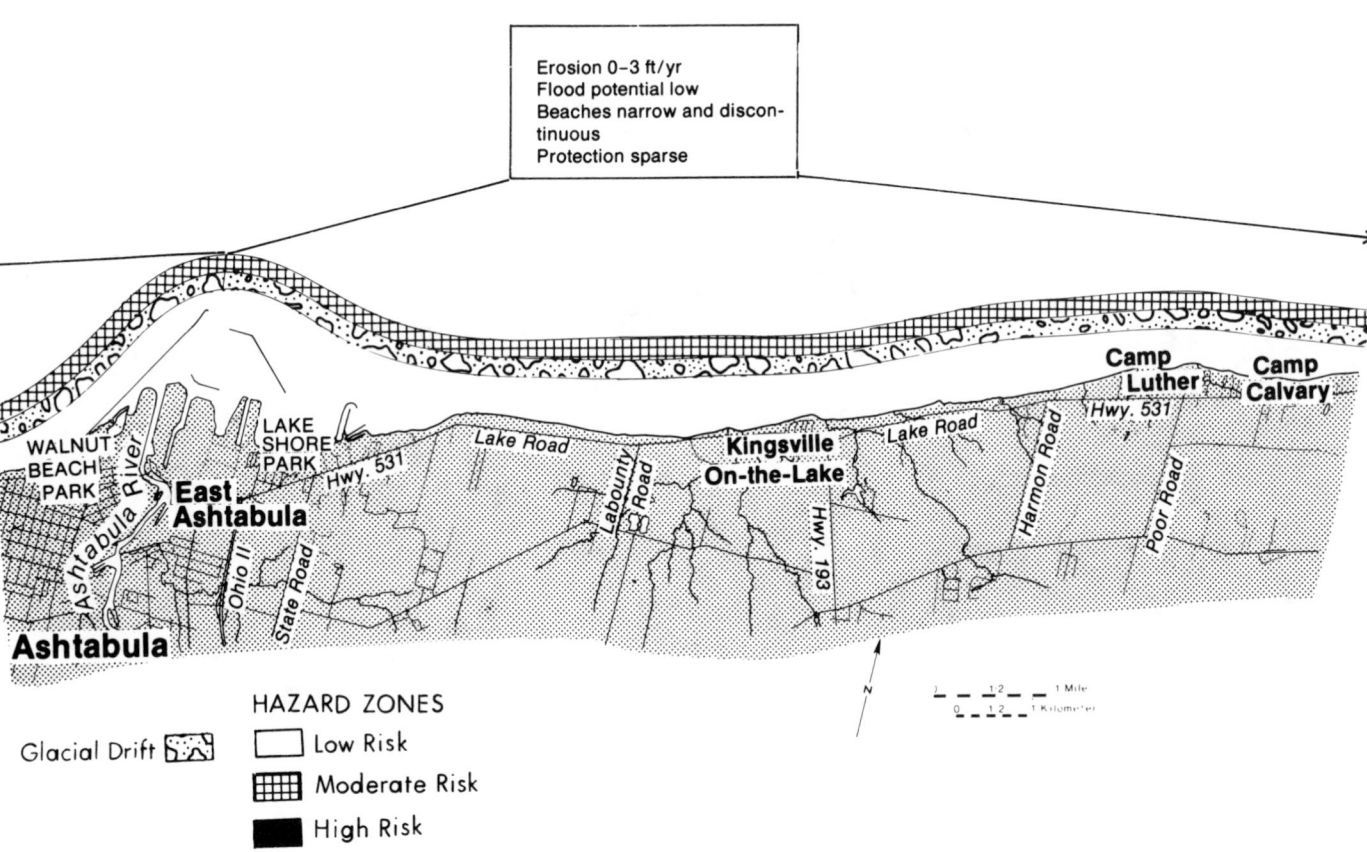

Erosion 0–3 ft/yr
Flood potential low
Beaches narrow and discontinuous
Protection sparse

WALNUT BEACH PARK

LAKE SHORE PARK

Ashtabula River

East Ashtabula

Ashtabula

Hwy. 531

Ohio II

State Road

Lake Road

Labounty Road

Kingsville On-the-Lake

Lake Road

Hwy. 193

Harmon Road

Poor Road

Hwy. 531

Camp Luther

Camp Calvary

N

1 — 1 2 — 1 Mile
0 — 1 2 — 1 Kilometer

HAZARD ZONES

Glacial Drift

☐ Low Risk

▦ Moderate Risk

■ High Risk

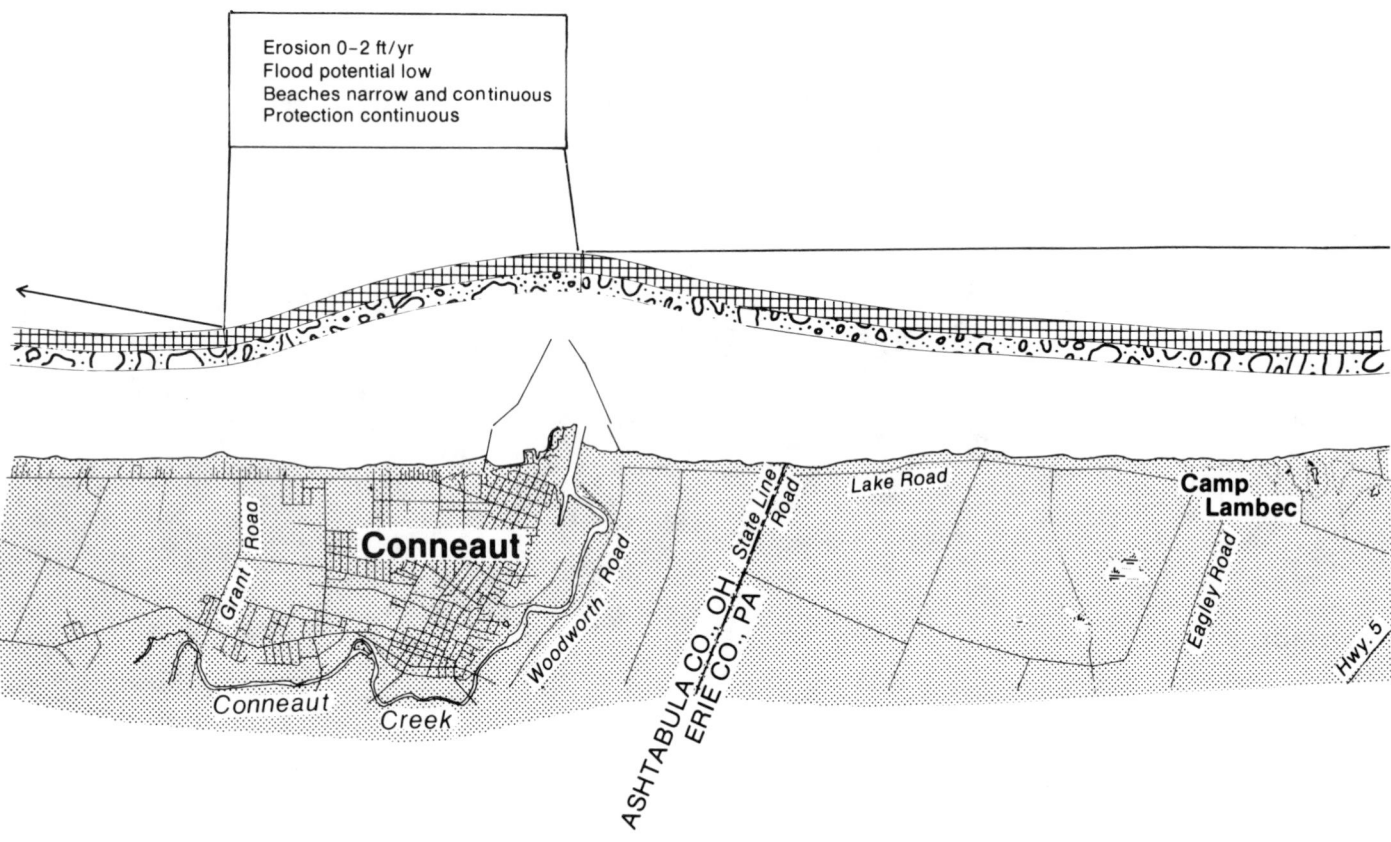

Erosion 0–2 ft/yr
Flood potential low
Beaches narrow and continuous
Protection continuous

Conneaut

Camp Lambec

Conneaut Creek

Grant Road

Woodworth Road

Lake Road

Eagley Road

Hwy. 5

ASHTABULA CO., OH — State Line Road
ERIE CO., PA

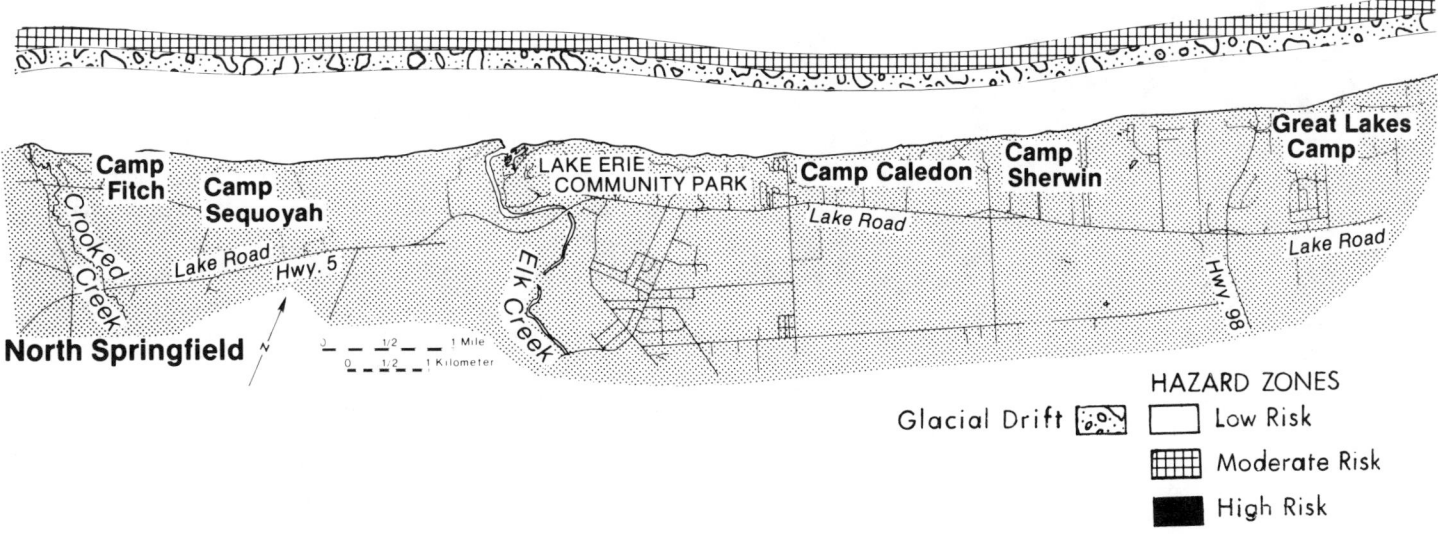

Erosion 0–3 ft/yr
Flood potential low except near stream mouths
Beaches narrow and discontinuous
Protection sparse

Camp Fitch
Camp Sequoyah
Crooked Creek
Lake Road
Hwy. 5
North Springfield
Elk Creek
LAKE ERIE COMMUNITY PARK
Camp Caledon
Lake Road
Camp Sherwin
Great Lakes Camp
Lake Road
Hwy. 98

0 ½ 1 Mile
0 ½ 1 Kilometer

HAZARD ZONES

Glacial Drift

☐ Low Risk
▦ Moderate Risk
■ High Risk

4.30 Site analysis: Camp Calvary to Great Lakes Camp

west end of the lake than it is for a 90-to-100-foot bluff along this stretch of shore. Flooding is not usually a problem, except when people build homes on the stream-mouth beaches, as at the mouth of Walnut Creek (Manchester Beach), which is asking for trouble.

Erie County, Pennsylvania

Stretch 9: Presque Isle (fig. 4.32). Presque Isle, a flying sand spit (a spit that extends offshore from a straight shoreline), is a popular state park of uncertain geologic origin (fig. 4.34). Many scientists believe that the spit may have formed when sand was deposited by waves refracted by a now-submerged cross-lake moraine. Successive episodes of deposition built up the spit. Whatever the origin, Presque Isle is a tourist mecca with a major problem: in spite of shore protection expenses that have totaled more than $20 million since 1824, the U.S. Army Corps of Engineers has been unable to stabilize the spit permanently. Coastal engineers, such as Denton Clark of the Corps' Buffalo district office, are not surprised at this, for this spit, with its easily erodible sand and exposed location, is a classic coastal engineering problem. All types of structures have been tried here including groins, seawalls, and breakwaters. The present solution is beach nourishment. Sand is trucked in and distributed on the beach. The sand, however, needs constant replenishment as it is transported away in large quantities by westerly wind-driven waves. Another method the Corps would like to try, but which

4.31   Ohio Route 531 shore protection. The highway was nearly lost to the lake in the 1970s. Landscaping and a seawall have helped to stabilize the shore and temporarily protect the highway from being destroyed by the lake. Note relatively natural shore in front of houses.

would cost about $30 million, is the deployment of a set of breakwaters to shelter Presque Isle's beaches from waves and thus greatly reduce the longshore transport of sand. This latter approach is being used more and more in coastal engineering with generally favorable results, but the federal government (ultimately the taxpayers) must be willing to put up two-thirds of the cost and the state (again the taxpayers), the remaining one-third, in order to attempt what is almost certainly not a final solution. This truly is a value judgment, and the ends (multiple ongoing engineering projects) may exhaust the means (future tax revenues needed to run such projects).

Erosion rates are highly variable with maximum rates of about eight feet per year. According to historic accounts the western (connected) end of the spit has migrated about three miles to the east. Sand transport rates are calculated by the Corps of Engineers to be about 280,000 cubic yards per year, a sizable quantity of sand for the Great Lakes. Much of today's problem may be tied to the major harbor jetties built in Ohio in the early 1800s. These structures, notably the ones at Cleveland, Fairport Harbor, Ashtabula, and Conneaut, have trapped sizable quantities of sand from the longshore transport system, thus depriving the downdrift shore of sand to the east of these structures. For example, the Fairport Harbor jetties are estimated to be trapping about 100,000 cubic yards of sand per year from the longshore transport system. One might argue that these structures may not be the sole cause of the problem, but it is almost certain that the reduced supply of sand from Ohio has contributed to accelerated erosion of Presque Isle, Pennsylvania.

The mainland shore of Erie, Pennsylvania, is characterized by fill land that fronts about two-thirds of the shore. The remainder of the shore consists of 90-foot bluffs and slopes of shale capped by till. There are no beaches because Presque Isle cuts off the sand moving along the shore from west to east, and little sand is eroded from the lakeshore. What sand there is, is transported offshore by short-period waves that reflect off the shale at the base of the bluffs. With the shale toe, and with Presque Isle acting as a giant breakwater, there is little erosion, and thus the area is quite stable. Erosion rates range from 0 to 2 feet per year. Because of the protected nature of the bay, and because of the extensive fill land, erosion of the original shore land should not be a problem along this segment of shore.

Reach 3 (U.S.): Erie, Pennsylvania (Presque Isle), to Buffalo, New York

This reach, along the southeast shore of Lake Erie (fig. 1.1), is characterized by moderate to high relief rock (shale) capped by till bluffs. Unlike at the west end of the lake, storm surges at this end of the lake cause little flooding because of the bluffs. However, relatively low-lying stretches and stream mouths are flooded during major surges at high lake level stages as in December 1985.

Continuous beaches front this shore except where interrupted by groins or jetties. The beaches along the Erie County shore of

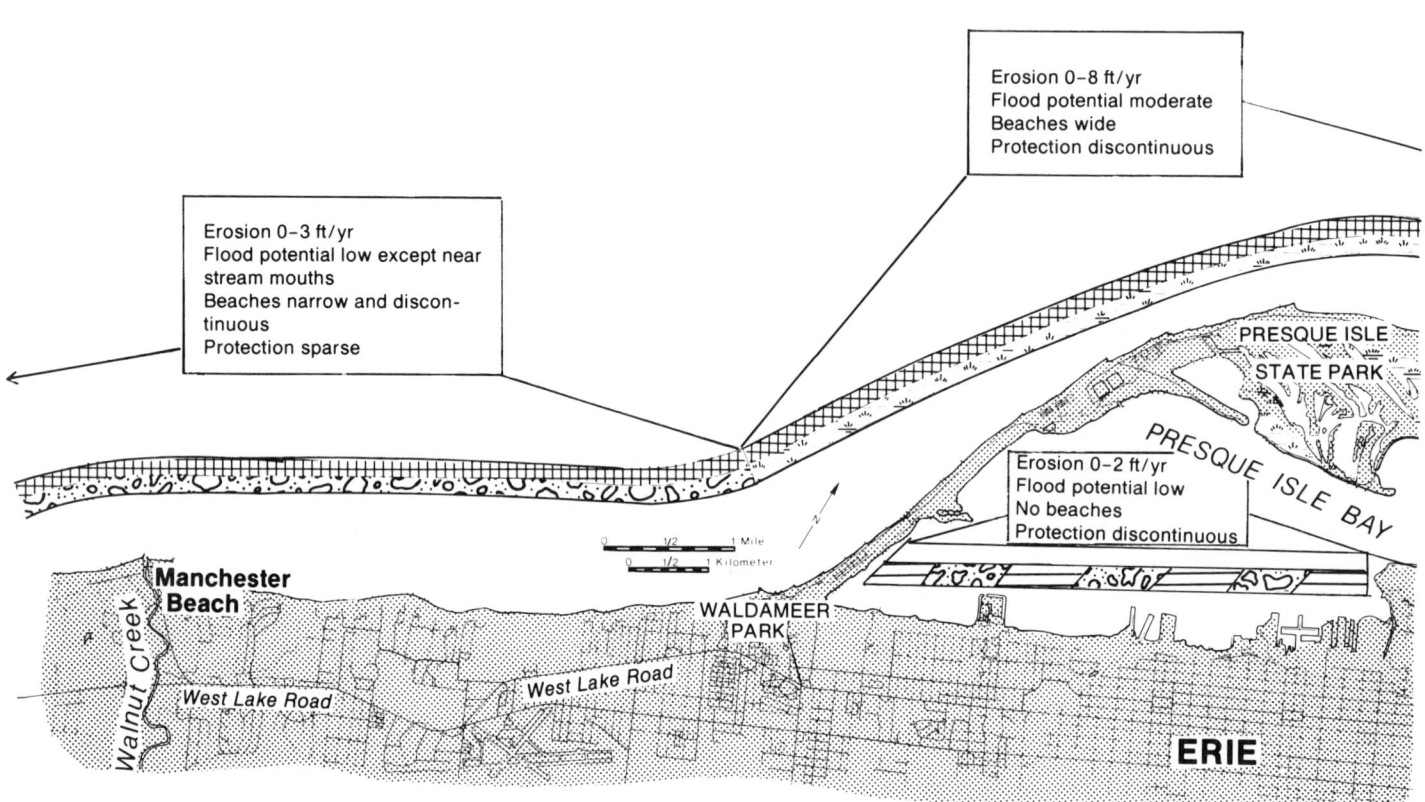

Erosion 0–8 ft/yr
Flood potential moderate
Beaches wide
Protection discontinuous

Erosion 0–3 ft/yr
Flood potential low except near
stream mouths
Beaches narrow and discon-
tinuous
Protection sparse

Erosion 0–2 ft/yr
Flood potential low
No beaches
Protection discontinuous

PRESQUE ISLE
STATE PARK

PRESQUE ISLE BAY

Manchester
Beach

Walnut Creek

West Lake Road

West Lake Road

WALDAMEER
PARK

ERIE

0 ½ 1 Mile
0 ½ 1 Kilometer

4.32 Site analysis: Great Lakes Camp to Twelvemile Creek

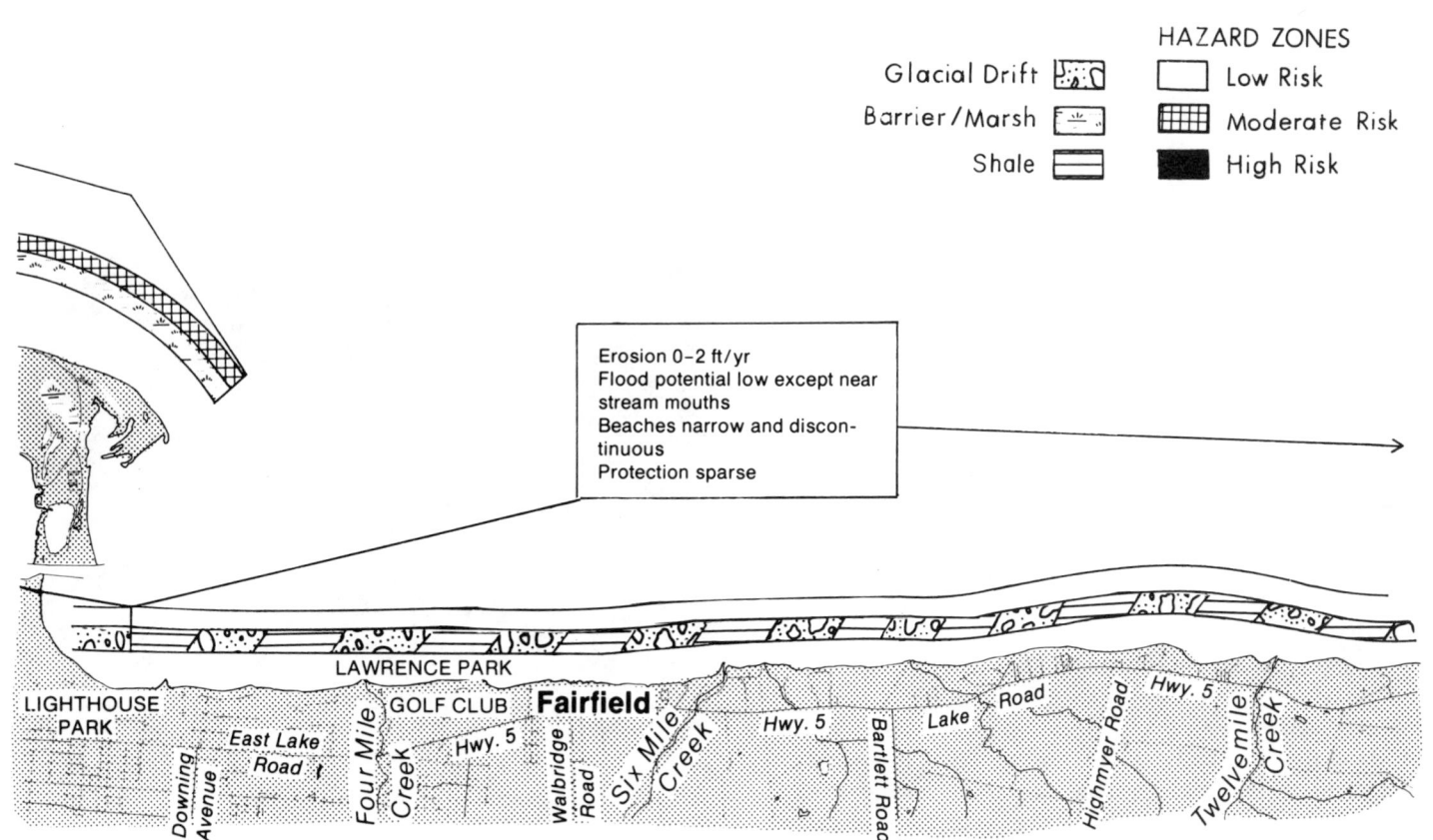

HAZARD ZONES

Glacial Drift

Barrier/Marsh

Shale

Low Risk

Moderate Risk

High Risk

Erosion 0–2 ft/yr
Flood potential low except near
stream mouths
Beaches narrow and discon-
tinuous
Protection sparse

LIGHTHOUSE
PARK

LAWRENCE PARK

GOLF CLUB **Fairfield**

Downing
Avenue

East Lake
Road

Four Mile
Creek

Hwy. 5

Walbridge
Road

Six Mile
Creek

Hwy. 5

Lake   Road

Bartlett Road

Highmyer Road

Hwy. 5

Twelvemile
Creek

New York are particularly wide and continuous because of sand derived from shore erosion as well as from the tributary streams. The beaches from Erie, Pennsylvania, to Cattaraugus Creek, New York, are largely confined to embayments (large pocket beaches) or to stream mouths such as Chautauqua or Canadaway creeks where sand-size sediment is transported to the lake and where the stream flow acts as a barrier to longshore transport. East of Cattaraugus Creek, the larger streams and the greater amount of till and sand in the shore deposits have created long, and sometimes wide beaches in spite of the shale bluffs.

The closer to Buffalo, the greater the number of groins and jetties to trap sand and segment beaches. Although the waves are the highest at the east end of the lake because of fetch and deep water, erosion rates are relatively low (generally less than one foot per year) because of the resistant shale that fronts most of the shore. Fortunately, where shale is not exposed, the beaches are wider because of the greater amounts of sand and gravel contributed by shore erosion. Erosion rates even along the till and sand shores are generally less than two feet per year. Naturally, the rates are highest if the shale is densely fractured or faulted or if the till or nonrock shore is fronted by only narrow or non-existent beaches.

Shore protection for much of this reach is unnecessary because of the low erosion rate of the shale. However, along the shore fronted by till or other unconsolidated material, as along the Sister creeks, New York, stretch, or along the shore where till caps the shale near lake level as along the Lake Erie State Park,

4.33 Conneaut, west of the jetties. These houses were built on a slumped area, but adequate toe protection in the form of beaches and seawalls has helped to stabilize the slide.

4.34 Presque Isle State Park, one of the premier parks on the lake. Presque Isle is a curved (flying) sand spit whose origin may be related to an underwater ridge that crosses the lake from Long Point. This spit is of additional interest because of the extensive beach nourishment that has been conducted here.

New York, protection of too-close homes may be needed to prevent erosion. The magnitude of the erosion problem in the unconsolidated deposits is largely related to the thickness of the deposit. If the deposit is as thick as 20 or 30 feet the problem is much greater because there is more of a downward force on the slope. That is, the higher the shore the greater the forces acting on the slope to cause failure. Fortunately, shale makes up the toe of the slope along much of this reach. Because shale is fairly resistant to weathering and erosion, it acts as a fairly good natural barrier to wave erosion and as a stable foundation for the overlying deposits. However, if the shale is no more than 10 to 15 feet above lake level, wave runup and wave splash during storms can reach and erode the toe of more easily erodible deposits, causing a steepening or undercutting of the slope and subsequent slope failure. In this situation a structure to protect the base of the unconsolidated material is necessary, but the idea is to shield the

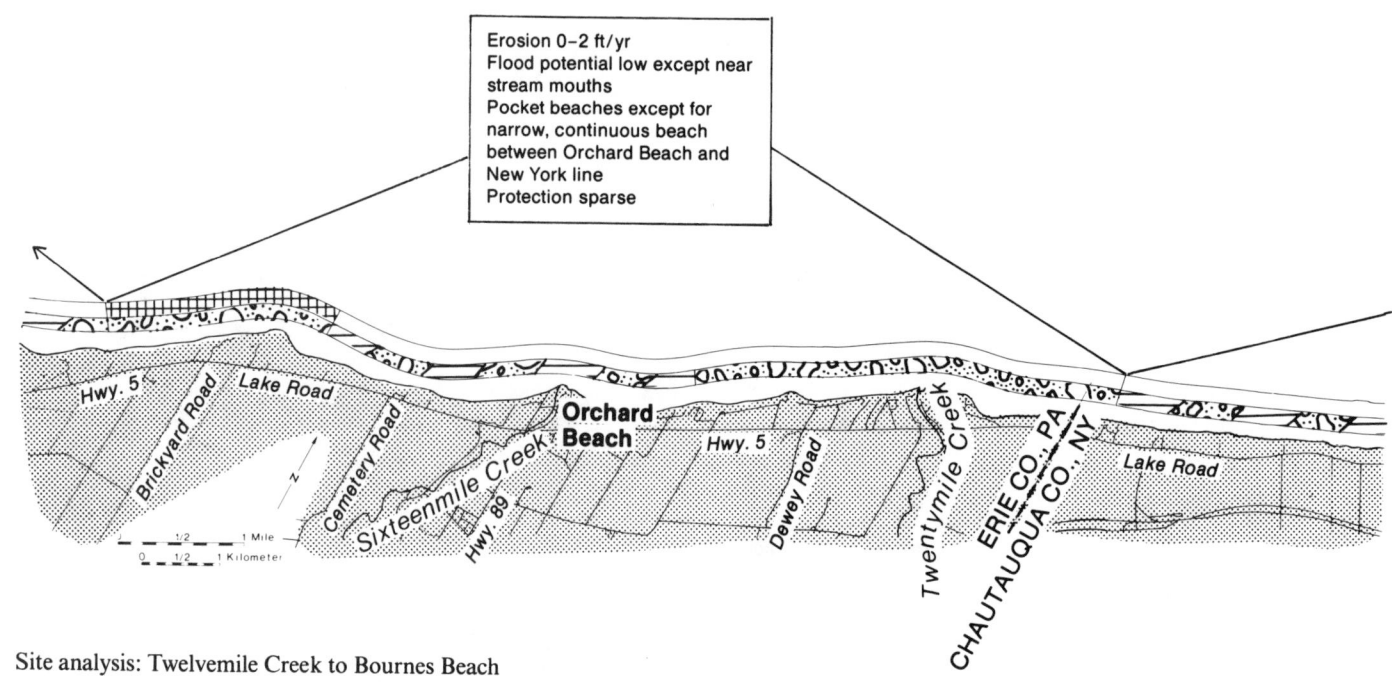

Erosion 0–2 ft/yr
Flood potential low except near
stream mouths
Pocket beaches except for
narrow, continuous beach
between Orchard Beach and
New York line
Protection sparse

4.35   Site analysis: Twelvemile Creek to Bournes Beach

shore from indirect wave forces (the waves have already broken
on the beach or shale below), and thus the seawall can be less
substantial than one built at lake level. In addition, grading or
terracing the slope plus planting a vegetative cover can increase
stability and reduce the erodibility of the slope. If the material

is not uniform and contains layers that trap and concentrate
groundwater, interceptor drains are a good idea. If there is an
erosion problem along a shore fronted mainly by unconsolidated
deposits, the procedures for stabilizing the shore are similar to
those for the Huron, Ohio, to Erie, Pennsylvania, reach. There

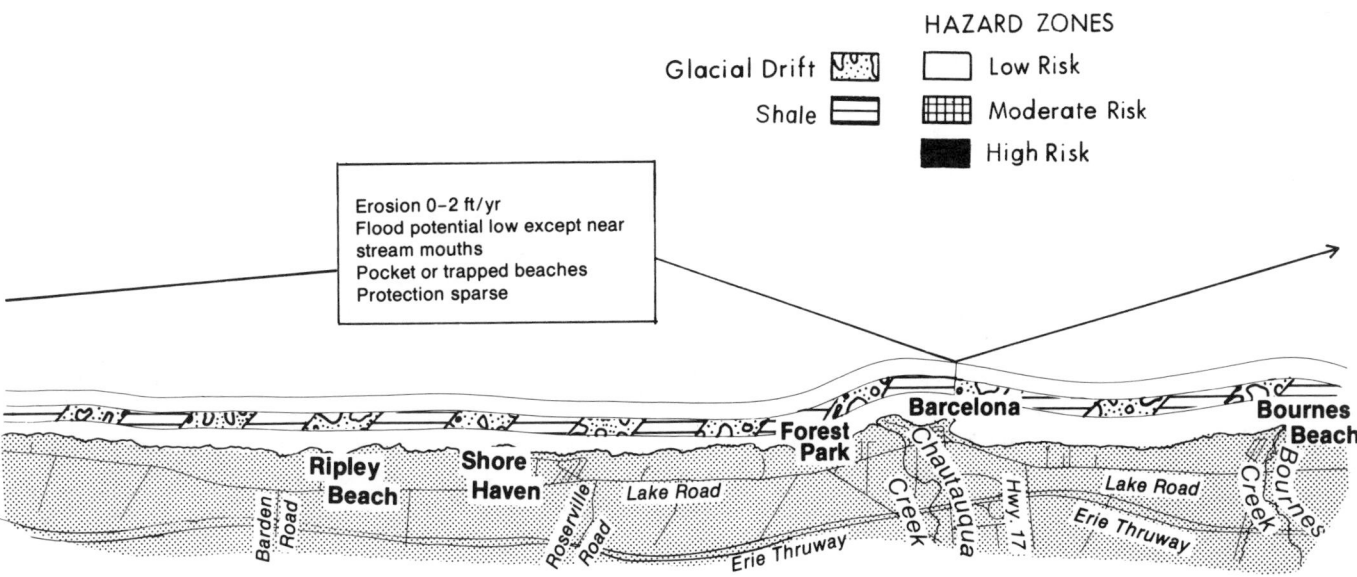

HAZARD ZONES

Glacial Drift ▨ ☐ Low Risk

Shale ▤ ▦ Moderate Risk

■ High Risk

Erosion 0–2 ft/yr
Flood potential low except near
stream mouths
Pocket or trapped beaches
Protection sparse

is, however, greater wave energy along this shore because of greater fetch and deeper water. Protection efforts should be concentrated on stabilizing the toe of the slope. In the long run, however, the most effective strategy, one that will save both money and beaches, is to move structures back from the lake.

Erie County, Pennsylvania, and Chautauqua County, New York

Stretch 1: Erie, Pennsylvania (Lighthouse Park), to Chautauqua Creek, New York (figs. 4.32 and 4.35). This stretch is one of the most variable along the eastern Lake Erie shore. Relief (except for the stream floodplains) ranges from about 30 feet in the Twentymile Creek area to 170 feet between Twelvemile and Sixteenmile Creeks. In general, the relief is 50 to 70 feet to the west of the Twelvemile-Sixteenmile Creek area and 30 to 50 feet to the east of this area.

Shale capped by till (and in places silt and sand, which provide an excellent soil for the local vineyards) is the typical bluff material. In the segment between Twelvemile Creek and the Pennsylvania-New York line, sections of the shore are composed solely of till and postglacial sediment. The shape of the shore provides a good record of the nature of the shore deposits: the shore fronted by till is nearly straight; the shore fronted by shale is irregular. The gentler slopes, greater erodibility of the till and postglacial sediment, and several small streams that hinder the west-to-east longshore transport of sand and gravel have led to the formation of a nearly continuous, narrow beach between Sixteenmile Creek and the Pennsylvania-New York line. The remainder of the stretch, except for pocket beaches and beaches trapped at the mouths of streams by jetties and breakwaters, is largely devoid of beaches. This lack of beaches is due to the small amount of sand and coarser sediment that has been eroded and transported along the shore, and to wave reflection off the shale at the base of the bluffs. Overall, the shore is quite stable in terms of erosion. Although the till and overlying unconsolidated sediment is easily weathered and eroded, the shale at the toe of the slope is not, and it acts as a barrier to Lake Erie's waves and

to downslope movement (fig. 4.36). Because of this, and because many of the homes and structures along this part of the lakeshore are situated a good distance from the bluff edge, erosion is not much of a problem. However, when people begin to develop the stream floodplains where there are beaches and easy access to the lake, there will be problems. On these floodplains, the greatest hazard is flooding, which threatens the small homes and structures.

Erosion rates range from 0 to 2 feet per year. Overall, the rates are highest where the shore is composed mainly of till. The toes of slopes composed of till are easily eroded by waves, increasing slope instability by increasing the slope angle. Water infiltrating through the deposits overlying the shale can create a slip plane between the shale and the till, which leads to slope failure. The shale by itself is good shore protection, although it does erode on the average about one foot per year.

Chautauqua County, New York

Stretch 2: Chautauqua Creek to Cattaraugus Creek (figs. 4.35, 4.37, and 4.38). This stretch, like the one to the west, is quite variable in relief and the nature of the shore deposits. Development here is more extensive, and the erosion threat has been met with the construction of several long seawalls. The relief ranges from about 10 feet at Van Buren Bay to about 100 feet at Slippery Rock Creek, a distance of only four miles down the shore. Overall the relief averages about 40 feet. Shale capped by till makes up the western half of the stretch. (There are short segments near Lake Erie State Park, for example, where till is exposed at lake level.) Shale, without till, makes up most of the eastern half (Dunkirk and east) of the stretch. Beaches, though discontinuous, front fairly long segments of the shore, notably in the areas of the Canadaway creeks and Cattaraugus Creek. In these areas sand and gravel is transported to the lake by the

4.36 Shale capped by till. The more resistant (but still erodible) shale keeps the shore from eroding as quickly as shore made up entirely of glacial deposits.

streams or is derived from erosion of the till. These beaches are essentially pocket beaches as they occur in embayments along the shore. Seawalls front developed segments, largely along Silver Creek-Hanford Bay, Dunkirk, and Van Buren Bay (fig. 4.39). Such structures typically front the more easily erodible shore materials of till and sand.

Erosion rates, as along the rest of the New York shore, range from 0 to 2 feet per year. The average is probably a little less than 1 foot per year. Again, the rates are highest where the more easily erodible till or postglacial sediment is exposed near lake level and near intensely fractured zones in the shale where the rock is more easily quarried by storm waves. Slope stability problems are largely confined to the western half of the stretch that has appreciable thicknesses of till overlying shale. Nonetheless, stability problems can occur along the entire stretch if the shale is sufficiently undercut to cause collapse. Houses should be set well back from the edge of the bluff.

Erie County, New York

Stretch 3: Cattaraugus Creek to Niagara River (United States-Canada border) (figs. 4.38 and 4.40). This stretch, because of its proximity to Buffalo and its relatively stable shore fronted by beaches, is the most heavily developed stretch along the New York shore. Relief ranges from about 10 feet along Evans Beach to about 120 feet on the west side of Eighteenmile Creek, and averages about 25 feet. Shale, capped in places by till, makes up

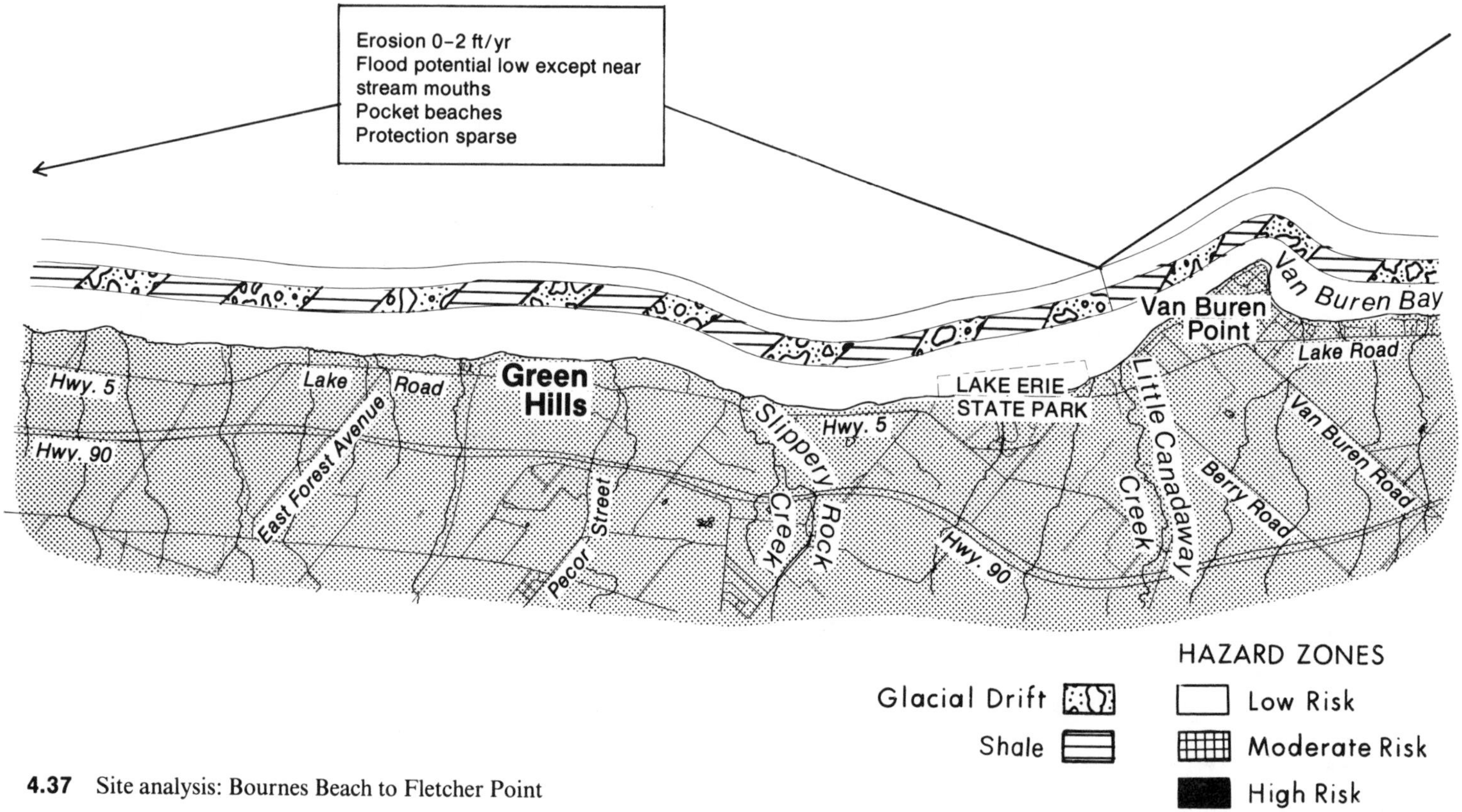

Erosion 0–2 ft/yr
Flood potential low except near stream mouths
Pocket beaches
Protection sparse

Hwy. 5

Hwy. 90

Lake Road

East Forest Avenue

Pecor Street

Green Hills

Slippery Rock Creek

Hwy. 5

Hwy. 90

LAKE ERIE STATE PARK

Little Canadaway Creek

Berry Road

Van Buren Point

Van Buren Bay

Lake Road

Van Buren Road

HAZARD ZONES

Glacial Drift

Shale

☐ Low Risk

▦ Moderate Risk

■ High Risk

4.37 Site analysis: Bournes Beach to Fletcher Point

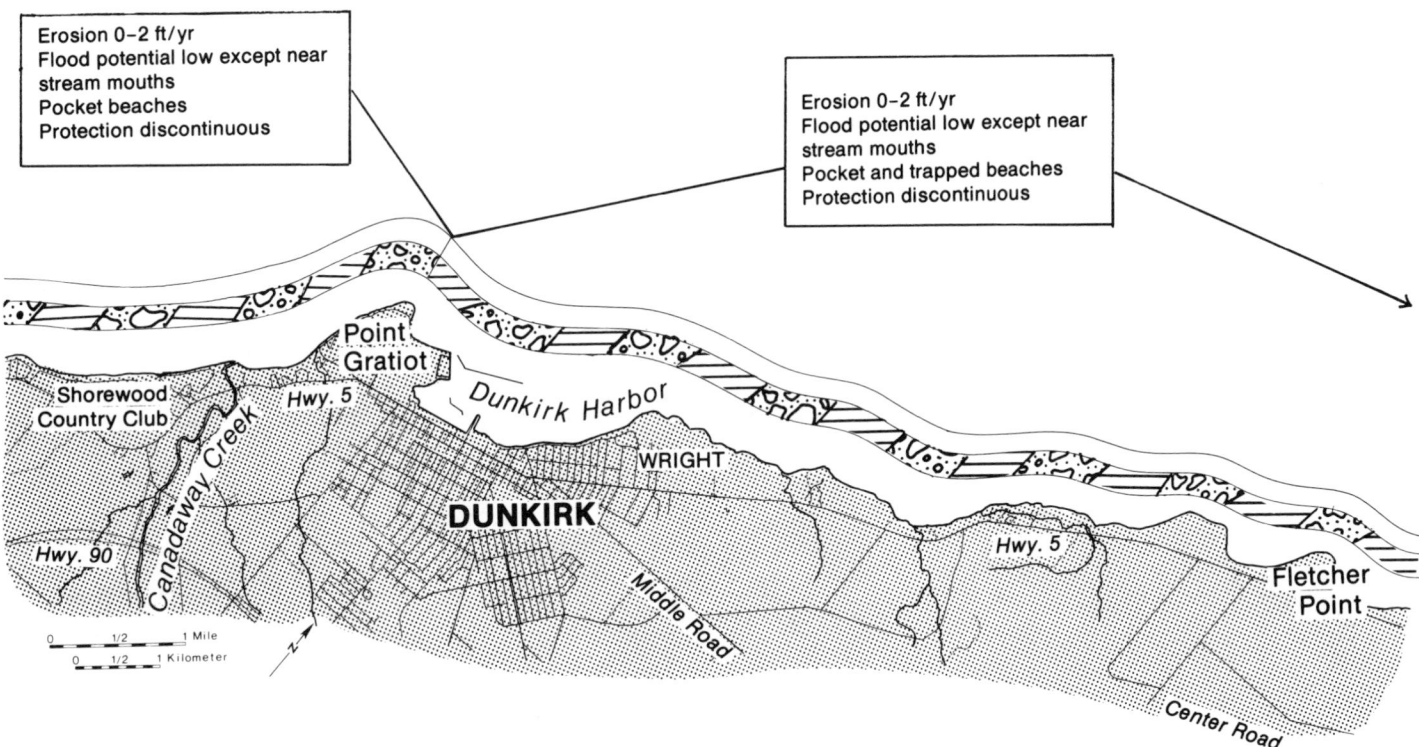

Erosion 0–2 ft/yr
Flood potential low except near
stream mouths
Pocket beaches
Protection discontinuous

Erosion 0–2 ft/yr
Flood potential low except near
stream mouths
Pocket and trapped beaches
Protection discontinuous

Point Gratiot

Dunkirk Harbor

WRIGHT

Shorewood Country Club

Hwy. 5

Canadaway Creek

DUNKIRK

Hwy. 90

Middle Road

Hwy. 5

Fletcher Point

Center Road

0   1/2   1 Mile
0   1/2   1 Kilometer

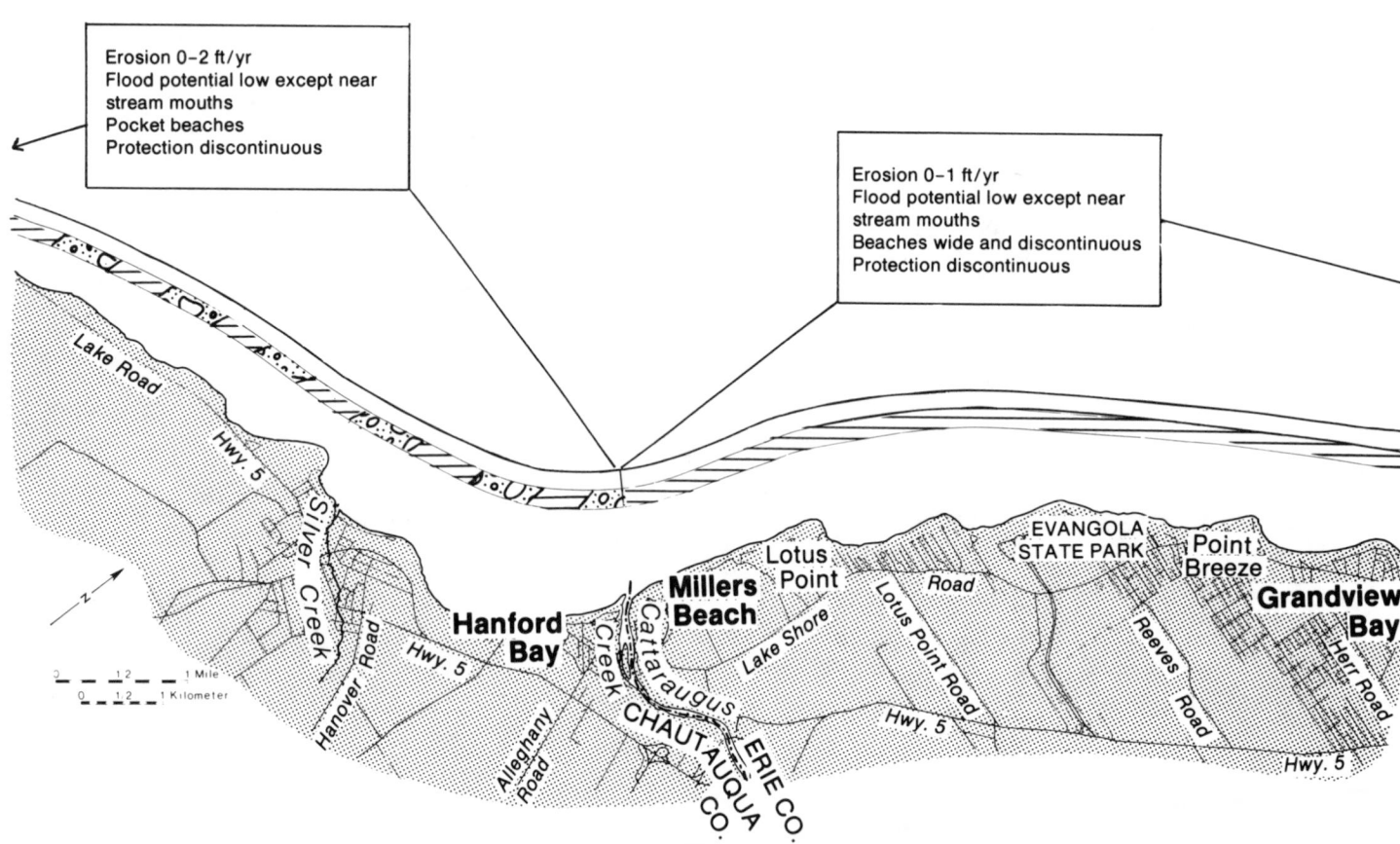

Erosion 0–2 ft/yr
Flood potential low except near
stream mouths
Pocket beaches
Protection discontinuous

Erosion 0–1 ft/yr
Flood potential low except near
stream mouths
Beaches wide and discontinuous
Protection discontinuous

Lake Road

Hwy. 5

Silver Creek

Hanover Road

Alleghany Road

N

0   1/2   1 Mile
0   1/2   1 Kilometer

Hanford Bay

Cattaraugus Creek

CHAUTAUQUA CO.

ERIE CO.

Millers Beach

Lotus Point

Lake Shore

Lotus Point Road

Hwy. 5

Road

EVANGOLA STATE PARK

Reeves Road

Point Breeze

Grandview Bay

Herr Road

Hwy. 5

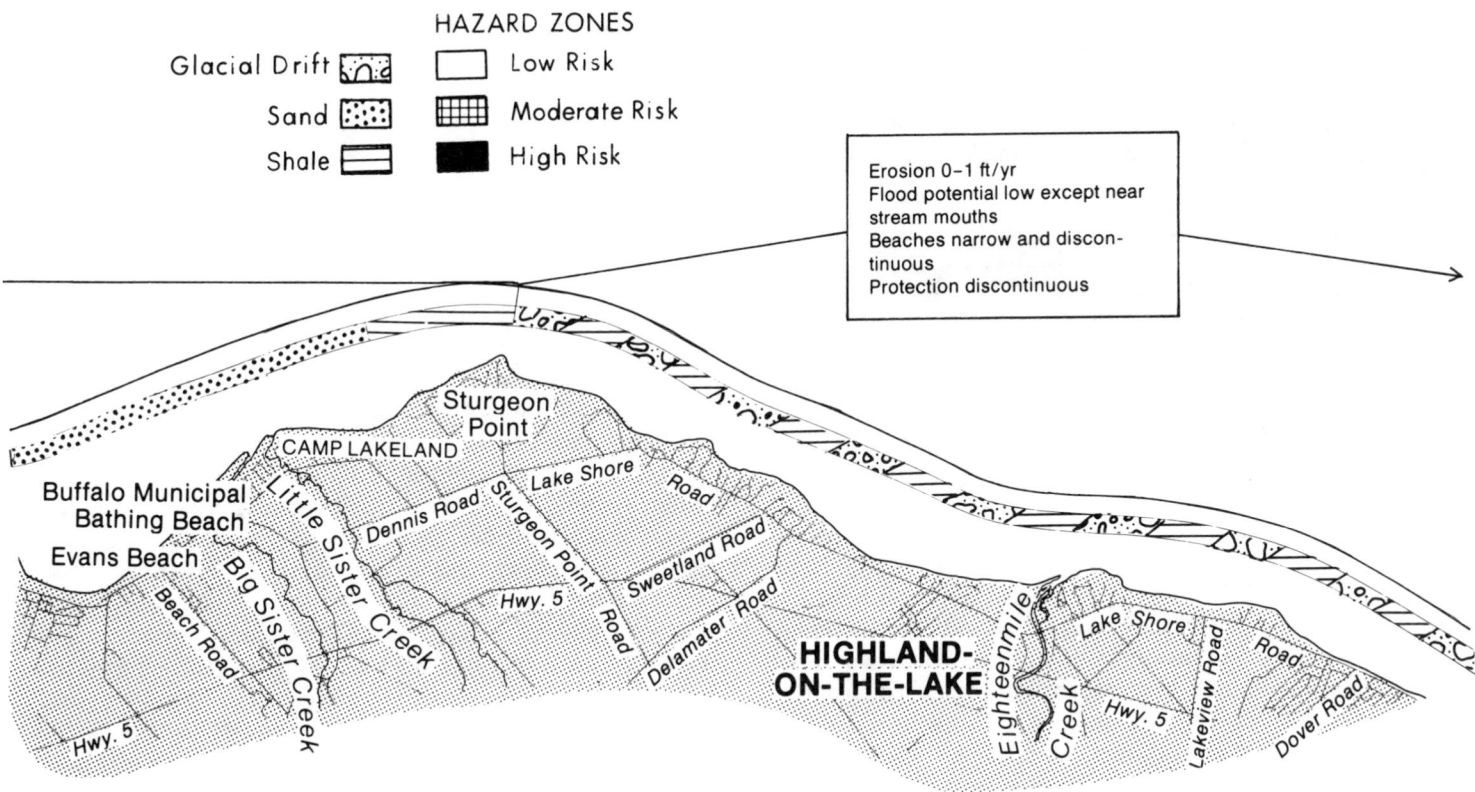

HAZARD ZONES

Glacial Drift — Low Risk

Sand — Moderate Risk

Shale — High Risk

Erosion 0–1 ft/yr
Flood potential low except near
stream mouths
Beaches narrow and discon-
tinuous
Protection discontinuous

Sturgeon Point

CAMP LAKELAND

Buffalo Municipal
Bathing Beach

Evans Beach

Little Sister Creek

Big Sister Creek

Beach Road

Dennis Road

Sturgeon Point Road

Lake Shore Road

Sweetland Road

Delamater Road

Hwy. 5

**HIGHLAND-
ON-THE-LAKE**

Eighteenmile Creek

Lake Shore Road

Lakeview Road

Dover Road

Hwy. 5

4.38 Site analysis: Fletcher Point to Dover Road

4.39 Van Buren Point area. A low relief area subject to erosion and flooding during extreme storm surges at the east end of the lake.

about two-thirds of the shore with the remainder largely fill land with minor sand bluffs. The widest, nearly continuous beaches front the bluffs between Cattaraugus Creek and Sturgeon Point. These beaches are broken only by prominent headlands such as Lotus Point and Point Breeze. Much of the beach sediment is derived from the tributary streams such as Cattaraugus and the Sister creeks. This coarse sediment is kept along the shore by the headlands, which cause waves to refract more directly onto the shore than along the shore, thus keeping the sand in these broad pockets. Nonetheless there is enough sand in the longshore sys-

tem that beaches do front some of headlands. Shore protection structures should be prohibited from segments such as these in order to protect and preserve the natural beaches. Narrow beaches are nearly continuous between Sturgeon Point and Bay View, whereas beaches largely related to fill activities are concentrated along the fill-land face between Bay View and the Buffalo Outer Harbor. Seawalls and groins, mostly close to Buffalo, front the shore in places. For the most part, these structures are located along the segments with the lowest relief and the most easily erodible deposits, as along Evans Beach (fig. 4.41) and the Sister Creeks, and in front of Athol Springs and Bay View (fig. 4.42).

Erosion rates here are the lowest along the Lake Erie shore at 0 to 1 foot per year. The rates are low in spite of the exposed location because of the beaches, which absorb much of the wave energy, and the resistant nature of the shale, which erodes at quite low rates. The greatest problems are along the short segments between Eighteenmile Creek and Bay View where till and postglacial deposits make up much of the bluff. In these areas there are slope stability problems because of wave erosion at the toe of the bluffs. Flooding, because of major storm surges generated by westerly winds, is a significant problem along the low-relief portions of this shore. The December 2, 1985, storm set a new surge record of 580.6 feet or about 10 feet above the long-term level of Lake Erie. And this level does not even take into account the 12-foot waves that accompanied the storm surge!

The Canadian portion of Lake Erie's north shore extends from the mouth of the Detroit River on the west to the head of the Niagara River on the east. Its total length of approximately 530 kilometers (329 miles), oriented in a prevailing southwest to northeast direction, lies entirely in the province of Ontario. Much of the north shore consists of shore bluffs undergoing active erosion. Over geologic time, geological processes have shaped these bluffs into distinctive, large, scalloplike indentations, with large sand spits (Point Pelee, Pointe aux Pins and Long Point) at the vertices. Thus the shore is divided into three physiographical regions often referred to as the western, central, and eastern basins. In recent times, however, this shore has been divided into five political regions. From west to east these regions are the counties of Essex, Kent, and Elgin; and the regional municipalities of Haldimand-Norfolk and Niagara (fig. 1.1).

Reach 4 (Canada): County of Essex, Ontario

The length of shoreline in this county is approximately 80 kilometers from the Detroit River on the west to Wheatley Harbour on the east (figs. 4.43 to 4.46).

Two distinct shore types of predominantly low bluff and barrier-sand beaches define this stretch of Lake Erie. Low-lying barrier beaches are in evidence throughout the entire stretch and are most prominent at the mouth of several major creeks (Big, Fox, Cedar, Sturgeon and Hillman) draining into the lake. Such barrier beaches make up the entire western and eastern shorelines of Point Pelee, extending eight kilometers southward into

the lake. The entire landform of Point Pelee, a modern spit built of sand eroded from shore bluffs both to the west and east, is a direct consequence of a unique pattern of sediment deposition arising from the interaction and merging of independent westerly and easterly nearshore sediment-laden currents (fig. 4.47). Beaches located west of the tip of Point Pelee, while dynamic in nature, may be considered stable over long time periods. However, beaches located east of the tip of Point Pelee are vulnerable to the onslaught of waves generated by storm winds blowing from the east over a 300-kilometer fetch. The devastation caused by these waves is directly proportional to the duration of the storm. During such a storm the barrier beaches are washed over and often breached, flooding the impounded marsh of Point Pelee National Park and the valuable reclaimed agricultural depression land behind. This situation is further aggravated during periods of extended high lake level (fig. 4.48).

A number of protective structures along this stretch of shore (i.e., seawalls, revetments, groins, and breakwaters, using a variety of materials and structural techniques) have been implemented with little, if any, long-term success. With the exception of armor stones weighing upward of two tonnes (1 tonne = 1.102 tons), all protective devices appear to be relegated to eventual failure (fig. 4.49).

About 20 kilometers of the Essex County shore are accentuated by bluffs composed predominantly of glacial till overlain by silty clay, with discontinuous beaches at the toe. They range in height from 20 meters at Colchester to 5 meters at Kingsville

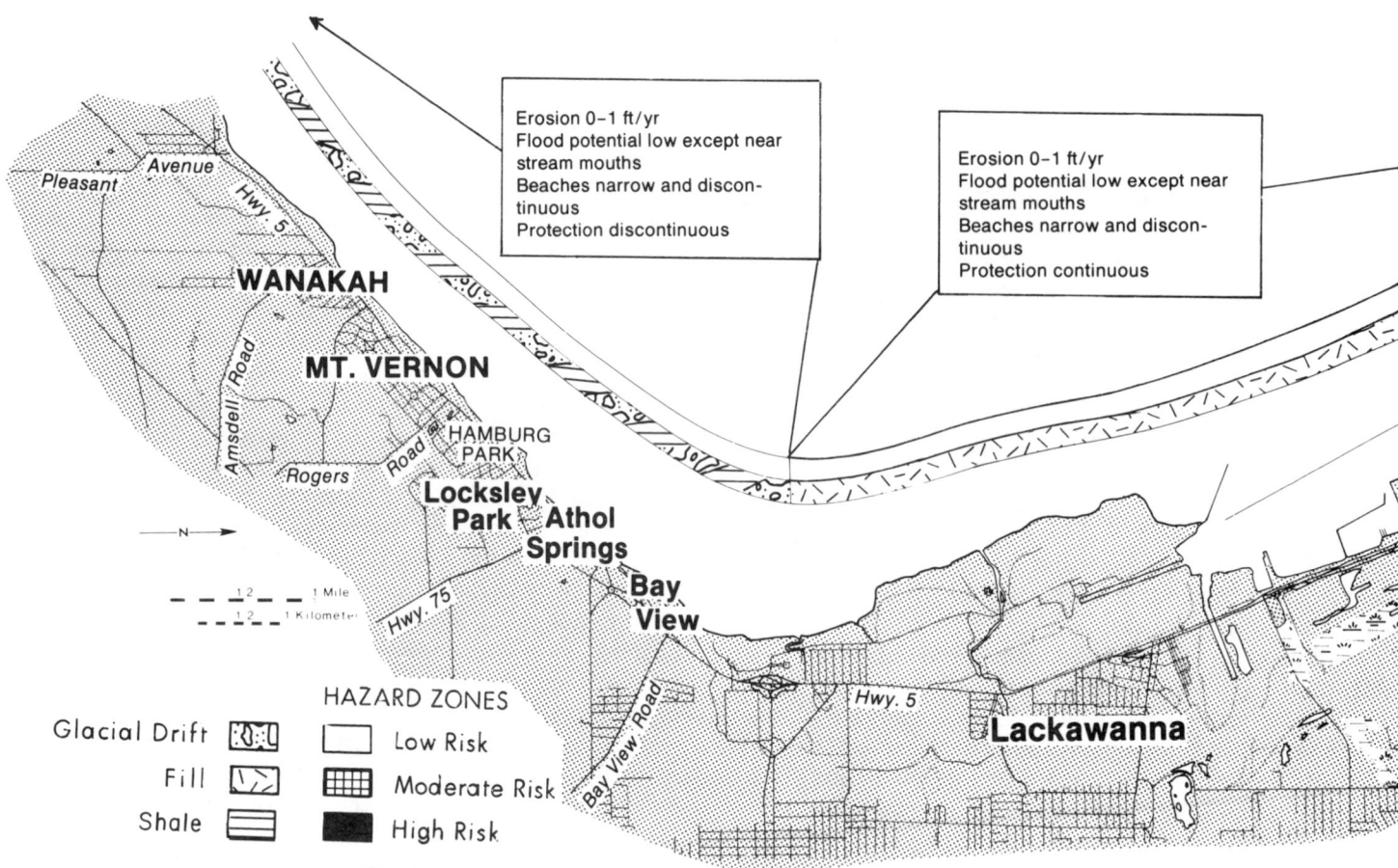

Erosion 0–1 ft/yr
Flood potential low except near
stream mouths
Beaches narrow and discon-
tinuous
Protection discontinuous

Erosion 0–1 ft/yr
Flood potential low except near
stream mouths
Beaches narrow and discon-
tinuous
Protection continuous

Pleasant   *Avenue*

Hwy. 5

WANAKAH

Amsdell Road

MT. VERNON

HAMBURG
PARK

Rogers   Road

**Locksley
Park**   **Athol
Springs**

**Bay
View**

Hwy. 75

Hwy. 5

Bay View Road

Lackawanna

N→

1 2   1 Mile
1 2   1 Kilometer

HAZARD ZONES

Glacial Drift   Low Risk

Fill   Moderate Risk

Shale   High Risk

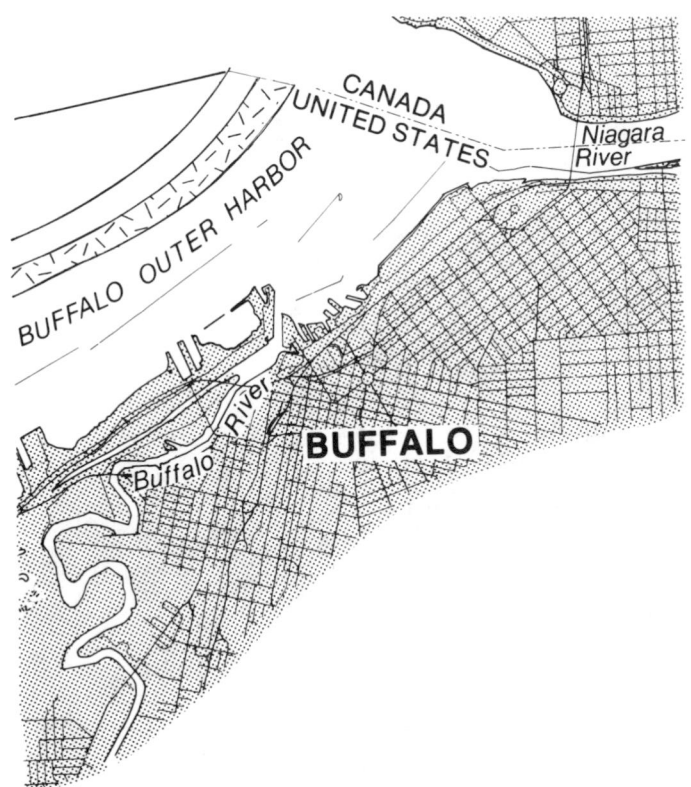

4.40 Site analysis: Dover Road to Buffalo

4.41 Evans Beach. This area is one of low relief, subject to erosion and flooding during extreme storm surges at the east end of the lake.
4.42 Bay View. This low relief area was badly damaged by the December 1985 storm surge (see fig. 1.7).

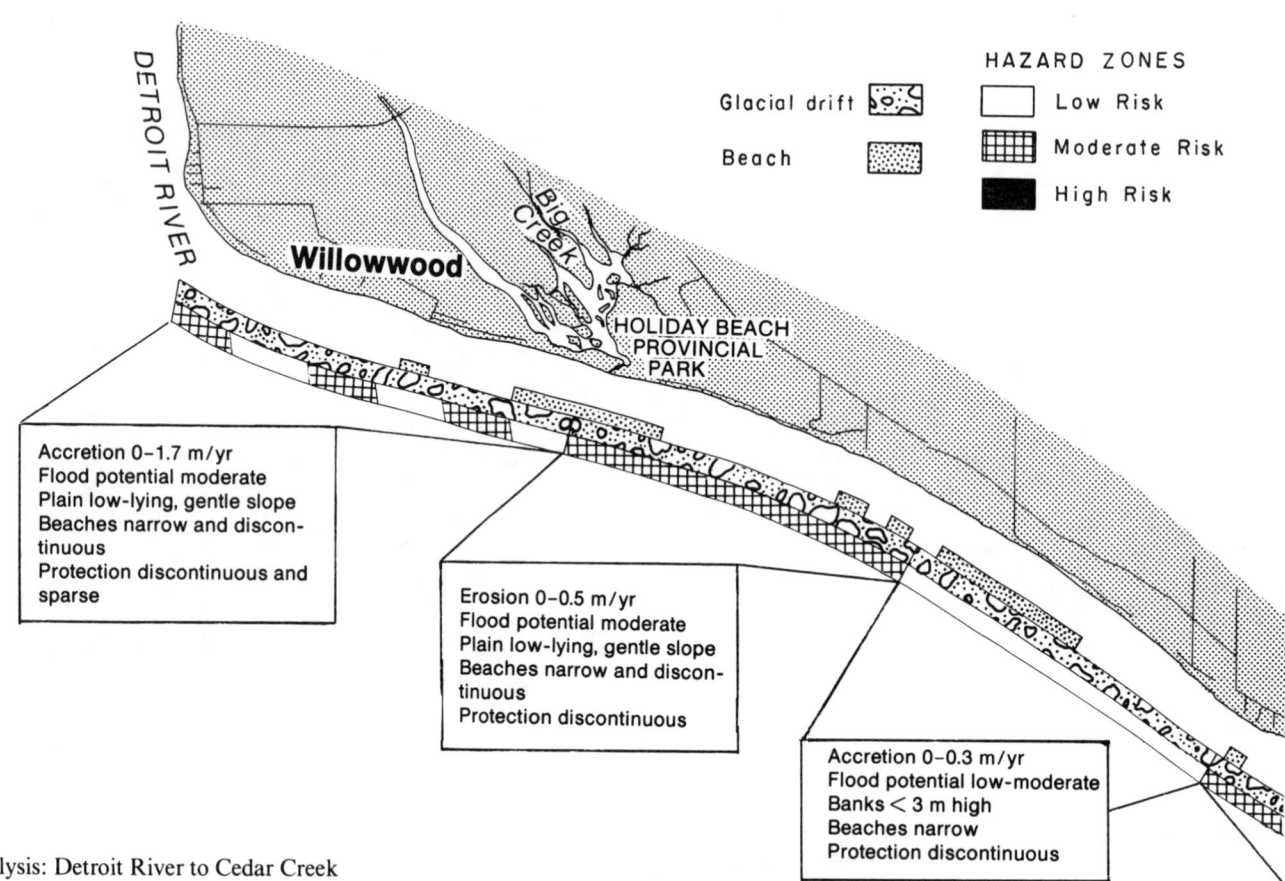

HAZARD ZONES

Glacial drift

Beach

Low Risk

Moderate Risk

High Risk

DETROIT RIVER

Willowwood

Big Creek

HOLIDAY BEACH
PROVINCIAL
PARK

Accretion 0–1.7 m/yr
Flood potential moderate
Plain low-lying, gentle slope
Beaches narrow and discon-
tinuous
Protection discontinuous and
sparse

Erosion 0–0.5 m/yr
Flood potential moderate
Plain low-lying, gentle slope
Beaches narrow and discon-
tinuous
Protection discontinuous

Accretion 0–0.3 m/yr
Flood potential low-moderate
Banks < 3 m high
Beaches narrow
Protection discontinuous

4.43   Site analysis: Detroit River to Cedar Creek

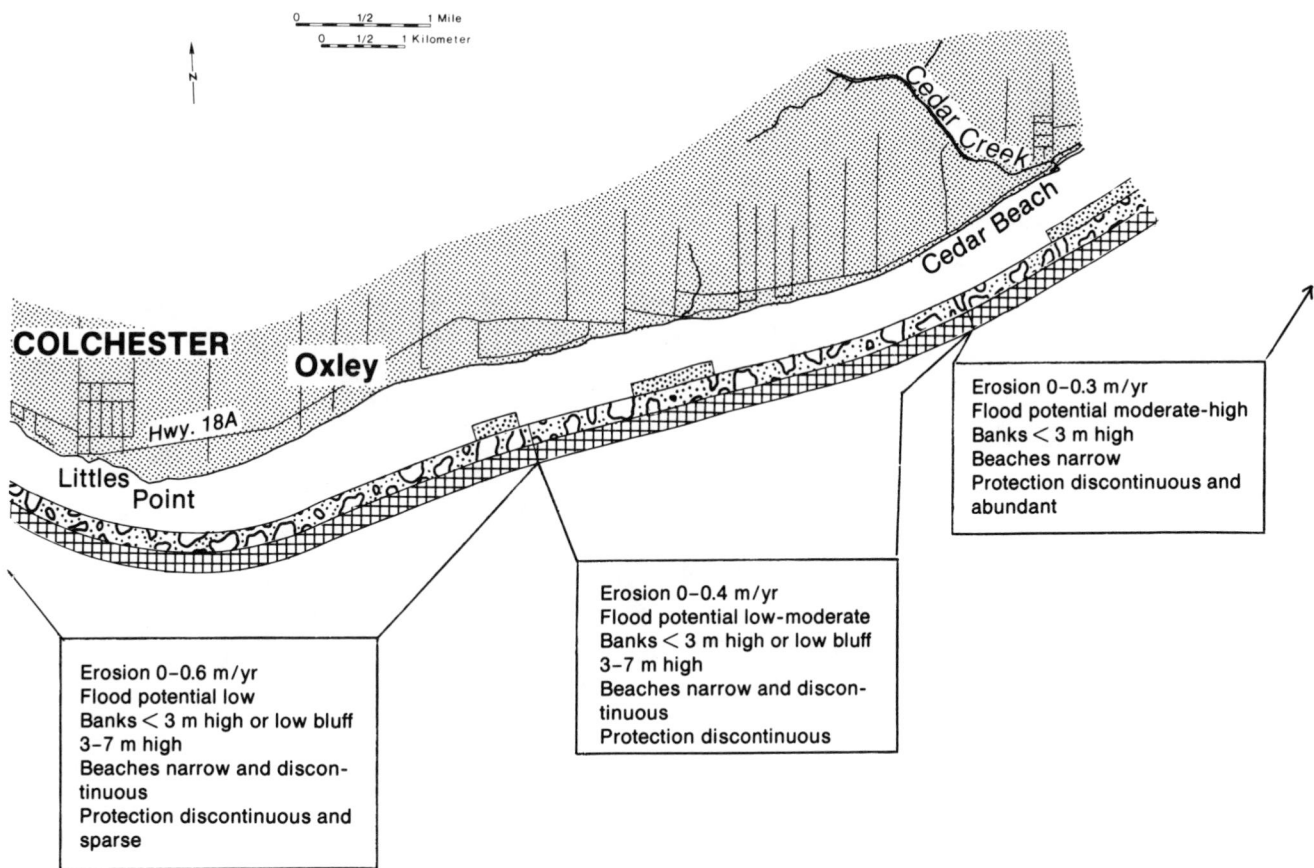

0   1/2   1 Mile

0   1/2   1 Kilometer

N

Cedar Creek

Cedar Beach

COLCHESTER   **Oxley**

Hwy. 18A

Littles Point

Erosion 0–0.3 m/yr
Flood potential moderate-high
Banks < 3 m high
Beaches narrow
Protection discontinuous and
abundant

Erosion 0–0.4 m/yr
Flood potential low-moderate
Banks < 3 m high or low bluff
3–7 m high
Beaches narrow and discon-
tinuous
Protection discontinuous

Erosion 0–0.6 m/yr
Flood potential low
Banks < 3 m high or low bluff
3–7 m high
Beaches narrow and discon-
tinuous
Protection discontinuous and
sparse

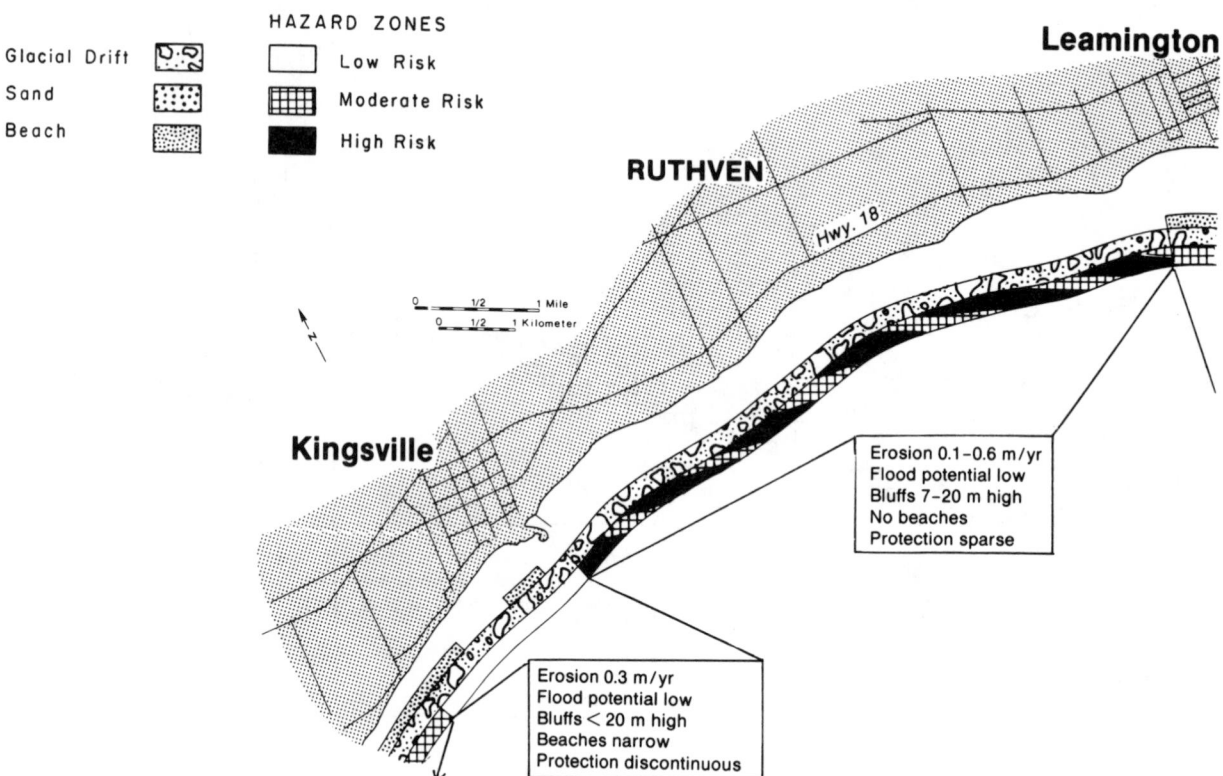

HAZARD ZONES

Glacial Drift

Sand

Beach

Low Risk

Moderate Risk

High Risk

Leamington

RUTHVEN

Hwy. 18

Kingsville

Erosion 0.1–0.6 m/yr
Flood potential low
Bluffs 7–20 m high
No beaches
Protection sparse

Erosion 0.3 m/yr
Flood potential low
Bluffs < 20 m high
Beaches narrow
Protection discontinuous

0   1/2   1 Mile
0   1/2   1 Kilometer

4.44   Site analysis: Kingsville to Point Pelee

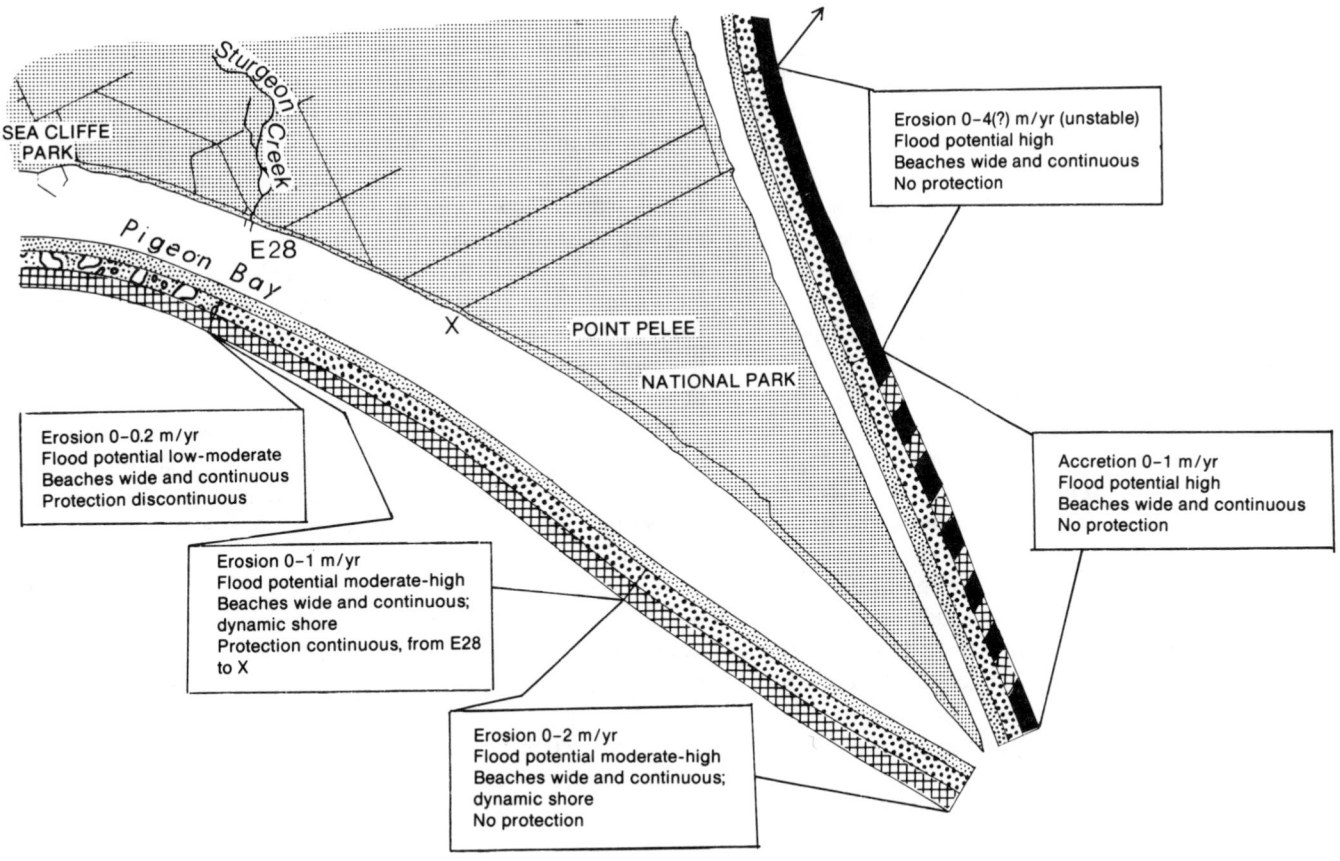

SEA CLIFFE PARK

Sturgeon Creek

E28

Pigeon Bay

X

POINT PELEE

NATIONAL PARK

Erosion 0-4(?) m/yr (unstable)
Flood potential high
Beaches wide and continuous
No protection

Accretion 0-1 m/yr
Flood potential high
Beaches wide and continuous
No protection

Erosion 0-0.2 m/yr
Flood potential low-moderate
Beaches wide and continuous
Protection discontinuous

Erosion 0-1 m/yr
Flood potential moderate-high
Beaches wide and continuous;
dynamic shore
Protection continuous, from E28
to X

Erosion 0-2 m/yr
Flood potential moderate-high
Beaches wide and continuous;
dynamic shore
No protection

and 19 meters at Leamington (fig. 4.50).

The bluffs at Colchester are particularly vulnerable to wave erosion at the toe because of the structural weakness of the shore soils. Bluffs east of Kingsville frequently undergo circular slope failures associated with ground water seepage augmented by drainage from the numerous septic systems of the extensive residential development along this shore. The proximity of large urban areas (Windsor and Detroit) continues to influence the type and pattern of land use in this reach. Predominant recreational use is evident in the numerous country clubs, beach associations, summer youth or trailer camps, municipal, provincial, and national parks, and abundance of seasonal residences occupying the beaches and the lower relief shores west of Kingsville and between Sturgeon Creek and Point Pelee National Park east of Leamington (fig. 4.44). The higher bluff between Kingsville and Leamington has retained its agricultural character with much of its shore occupied by ancestral homes, orchards, and greenhouses. Many individually constructed shore protection structures, such as seawalls, revetments, and groins afford some safeguard against erosion, but are generally inadequate due to improper design, lack of coordinated planning, and the prohibitive cost of "adequate" structures. However, long-term recession of the edge of the bluff varies from little or no recession west of Colchester to greater than 0.5 meters per year in the Leamington area.

Detached from the mainland and separating the relatively shallow (10 meters) westerly basin from the rest of the lake is a chain of bedrock islands. The largest of them, Pelee Island (fig. 4.45), is located some 13 kilometers southwest from the tip of Point Pelee. It is quasi-rectangular in shape, extending 8 kilometers from north to south and less than 5 kilometers from east to west, having an area of just over 4,000 hectares (9,880 acres). Most of the shore of Pelee Island consists of sand beach backed by a thin line of low dunes, with the exception of the northwest and southeast corners of the island, where limestone ledges form the shore. The north shore of the island along North Bay is in a lee of Lighthouse Point that is protected with a series of sandbars. The southwest portion of the shore along South Bay is a low bluff ending in a sand and gravel spit extending south for about 2 kilometers.

Most of the central portion of the island is made up of artificially drained agricultural land. Since this land has an elevation barely above lake level, shoreline dike systems were required as protection against flooding, particularly during major easterly storms (fig. 4.51).

The predominant land use on the island is agricultural. Cottage development is concentrated along the eastern shore where gently sloping beaches attracted some 175 cottages, 30 of which were damaged and 15 destroyed during the storm of November 1972. Since the cottages are situated on the lake side of the dikes as well as below them, every major easterly storm takes its toll, each time adding to the cumulative cost of repair and loss (fig. 4.52).

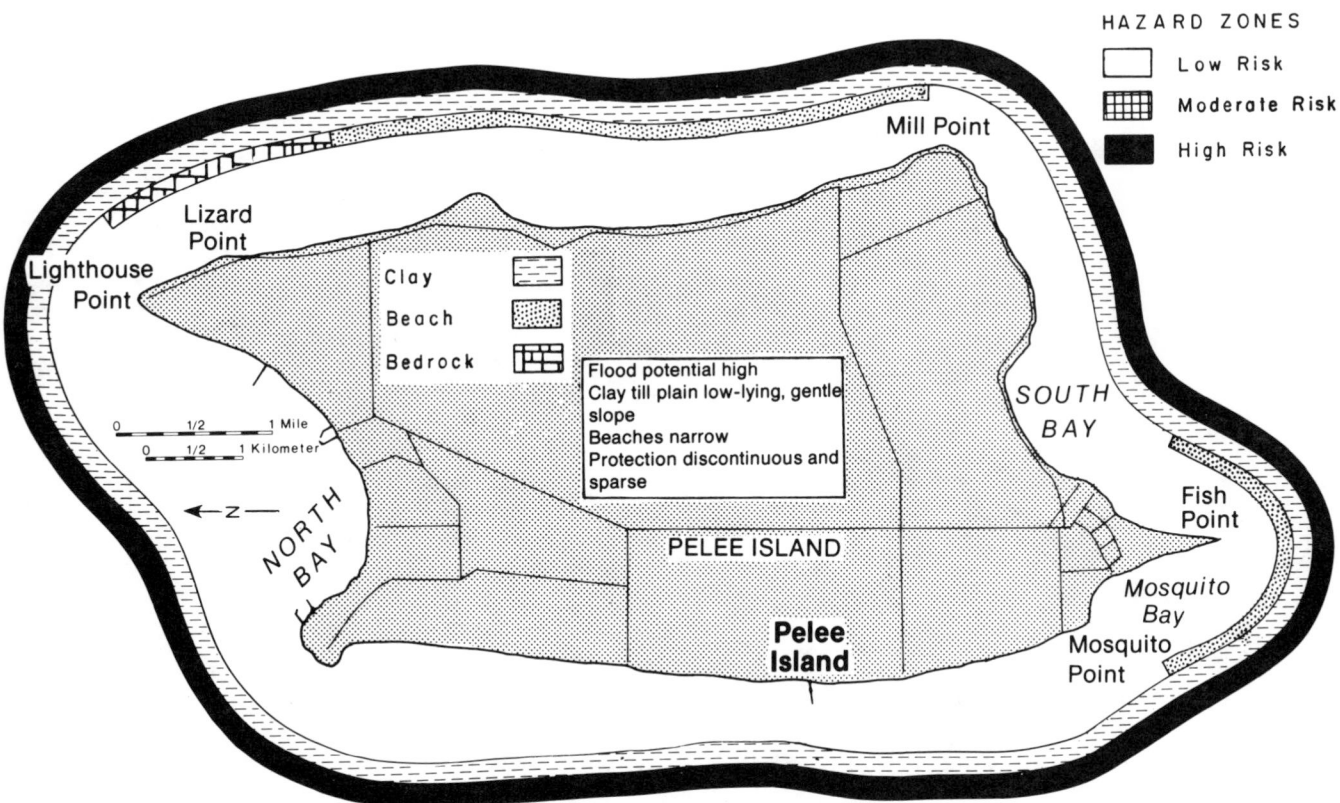

HAZARD ZONES

☐ Low Risk

▦ Moderate Risk

■ High Risk

Mill Point

Lizard Point

Lighthouse Point

Clay

Beach

Bedrock

Flood potential high
Clay till plain low-lying, gentle slope
Beaches narrow
Protection discontinuous and sparse

SOUTH BAY

0   1/2   1 Mile
0   1/2   1 Kilometer

←—z—

NORTH BAY

Fish Point

PELEE ISLAND

Mosquito Bay

Pelee Island

Mosquito Point

4.45 Site analysis: Pelee Island

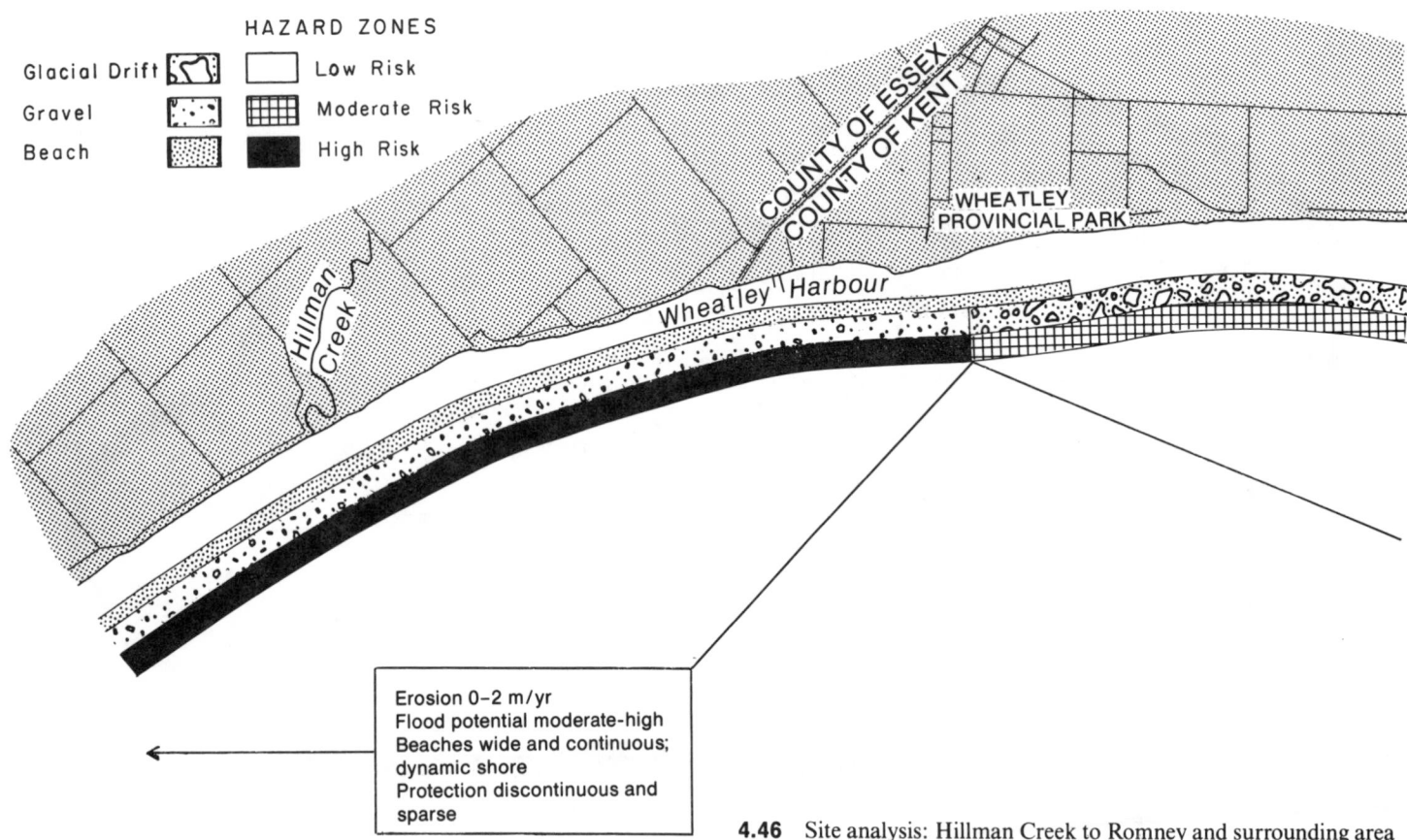

HAZARD ZONES

Glacial Drift ◇   Low Risk ▢

Gravel ▦   Moderate Risk ▦

Beach ▨   High Risk ■

COUNTY OF ESSEX
COUNTY OF KENT

WHEATLEY
PROVINCIAL PARK

Hillman Creek

Wheatley Harbour

Erosion 0–2 m/yr
Flood potential moderate-high
Beaches wide and continuous;
dynamic shore
Protection discontinuous and
sparse

4.46   Site analysis: Hillman Creek to Romney and surrounding area

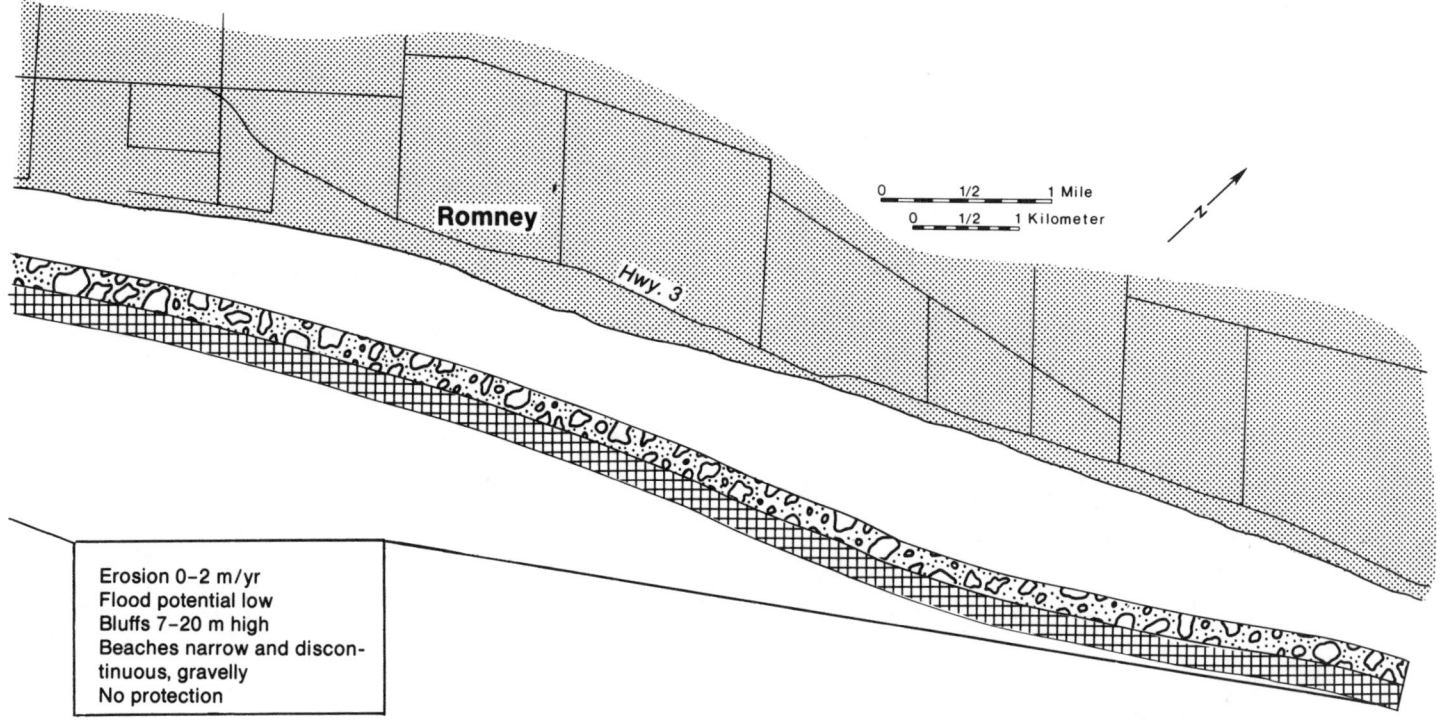

Romney

Hwy. 3

0   1/2   1 Mile

0   1/2   1 Kilometer

N

Erosion 0–2 m/yr
Flood potential low
Bluffs 7–20 m high
Beaches narrow and discon-
tinuous, gravelly
No protection

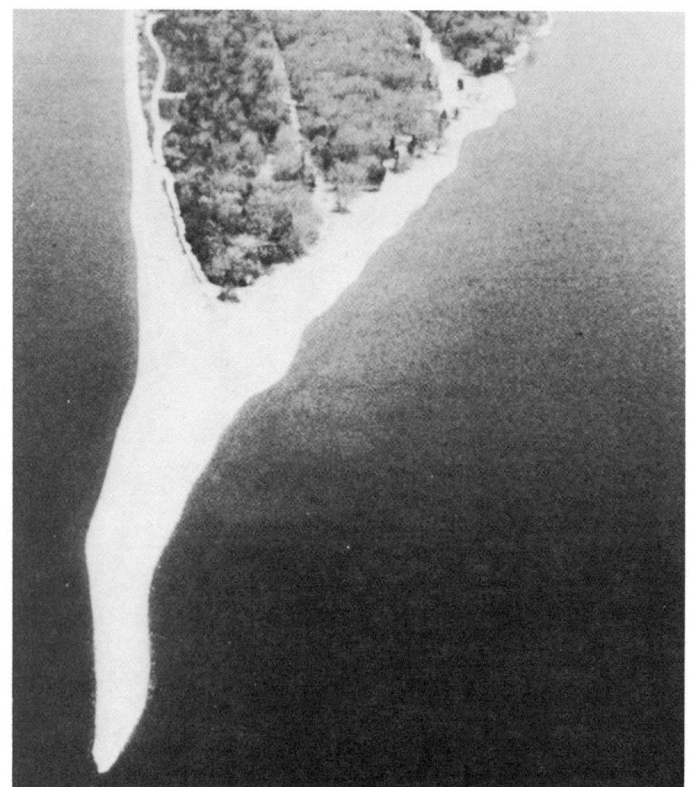

4.47   Oblique view looking north at sand spit of Point Pelee National Park, which is the most southerly portion of the mainland in Canada.

4.48   Fragile nature of eastern barrier beaches of Point Pelee can be easily observed in this photo.

4.49   Size relation between armor stone, the cottage, and its owner.

4.51   Barn isolated by floodwaters in the central portion of Pelee Island during the November 1972 storm.

4.52   Result of 1972 storm wave onslaught on cottage located on the eastern beach of Point Pelee.

4.50   Typical shore characteristics and development in the vicinity east of Kingsville.

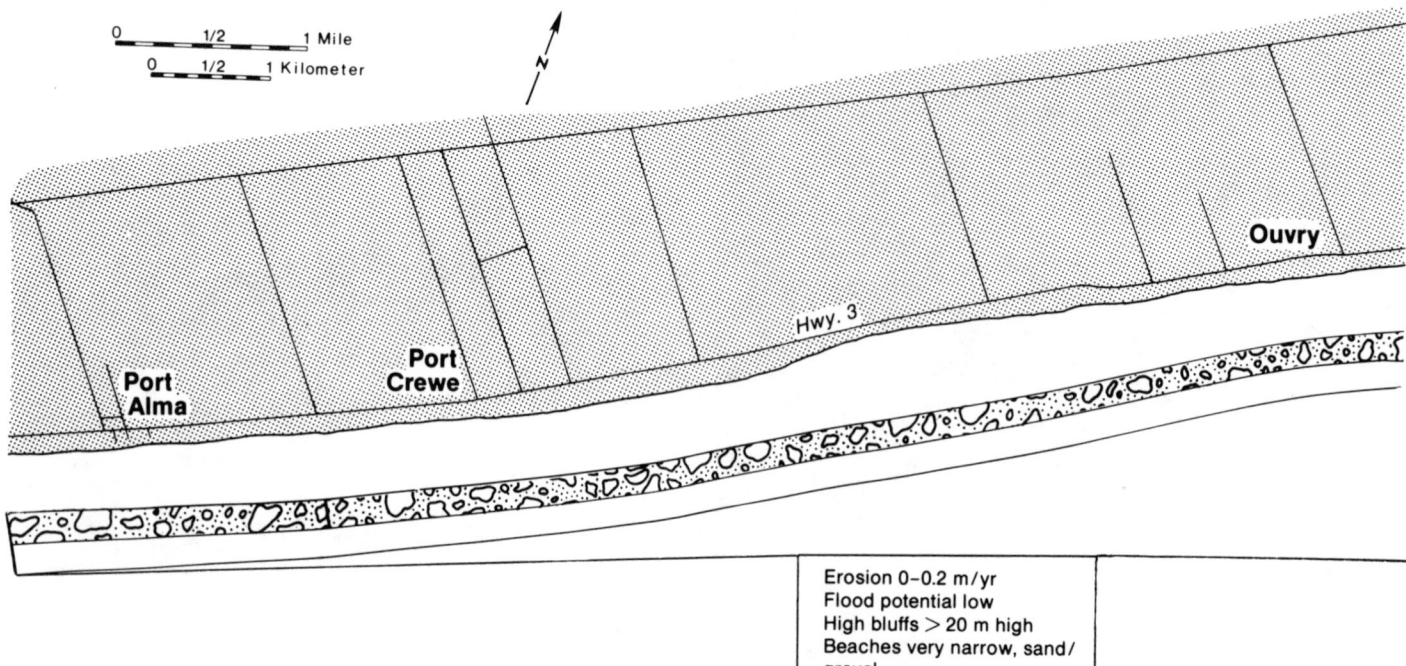

0 ─── 1/2 ─── 1 Mile
0 ─── 1/2 ─── 1 Kilometer

N

Ouvry

Hwy. 3

Port
Alma

Port
Crewe

Erosion 0-0.2 m/yr
Flood potential low
High bluffs > 20 m high
Beaches very narrow, sand/
gravel
No protection

4.53   Site analysis: Port Alma to Erie Beach

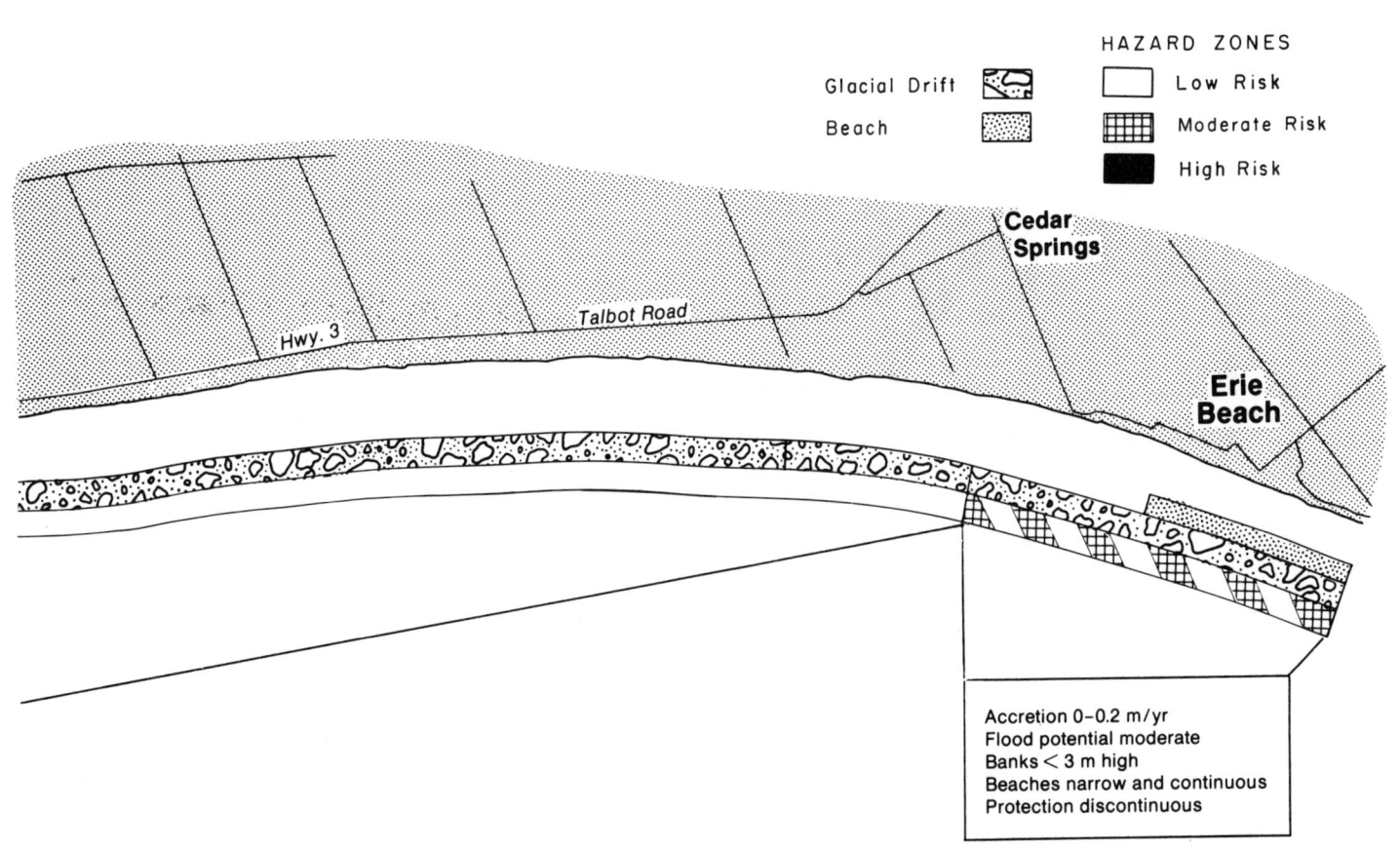

HAZARD ZONES

Glacial Drift

Beach

Low Risk

Moderate Risk

High Risk

Cedar
Springs

Erie
Beach

Hwy. 3

Talbot Road

Accretion 0–0.2 m/yr
Flood potential moderate
Banks < 3 m high
Beaches narrow and continuous
Protection discontinuous

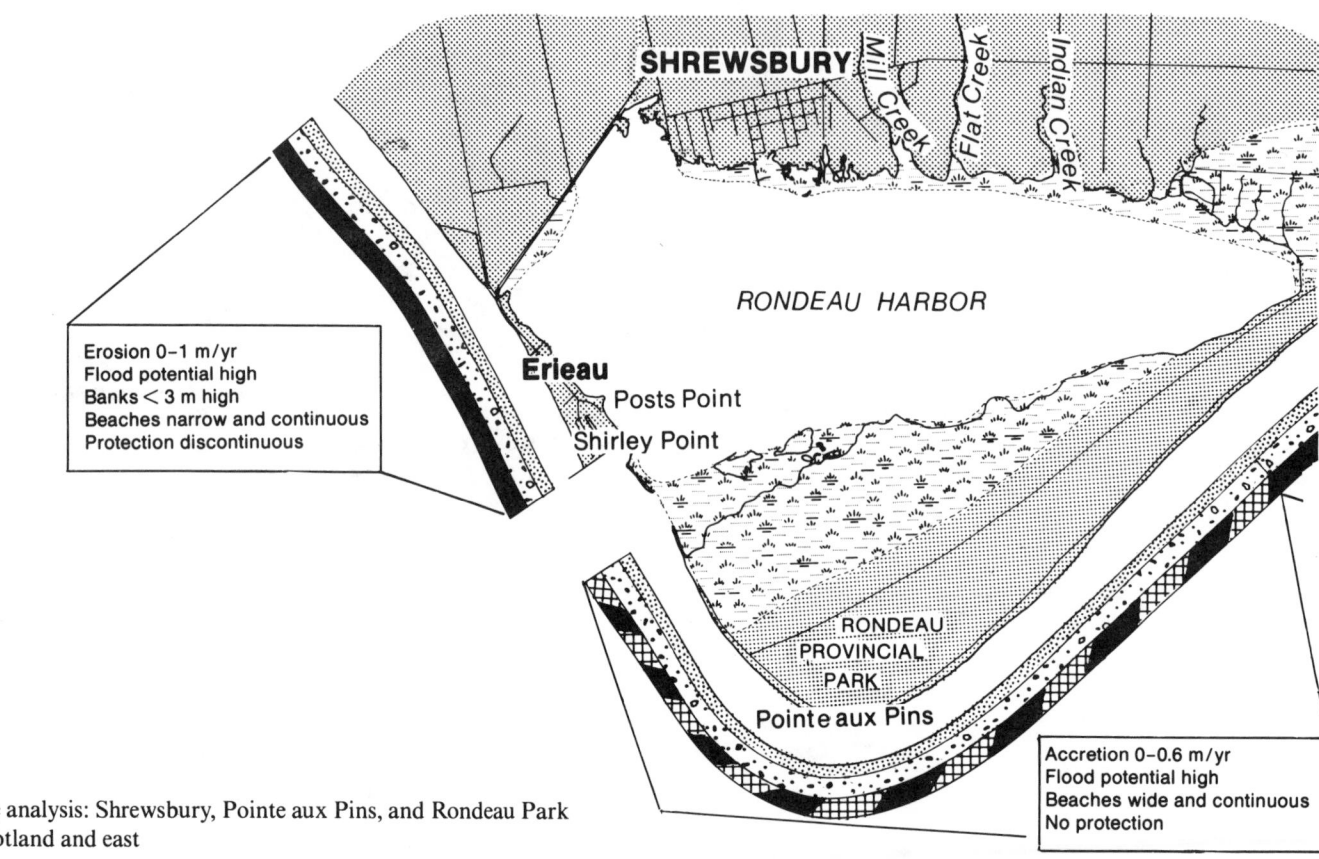

Erosion 0–1 m/yr
Flood potential high
Banks < 3 m high
Beaches narrow and continuous
Protection discontinuous

Accretion 0–0.6 m/yr
Flood potential high
Beaches wide and continuous
No protection

4.54 Site analysis: Shrewsbury, Pointe aux Pins, and Rondeau Park to New Scotland and east

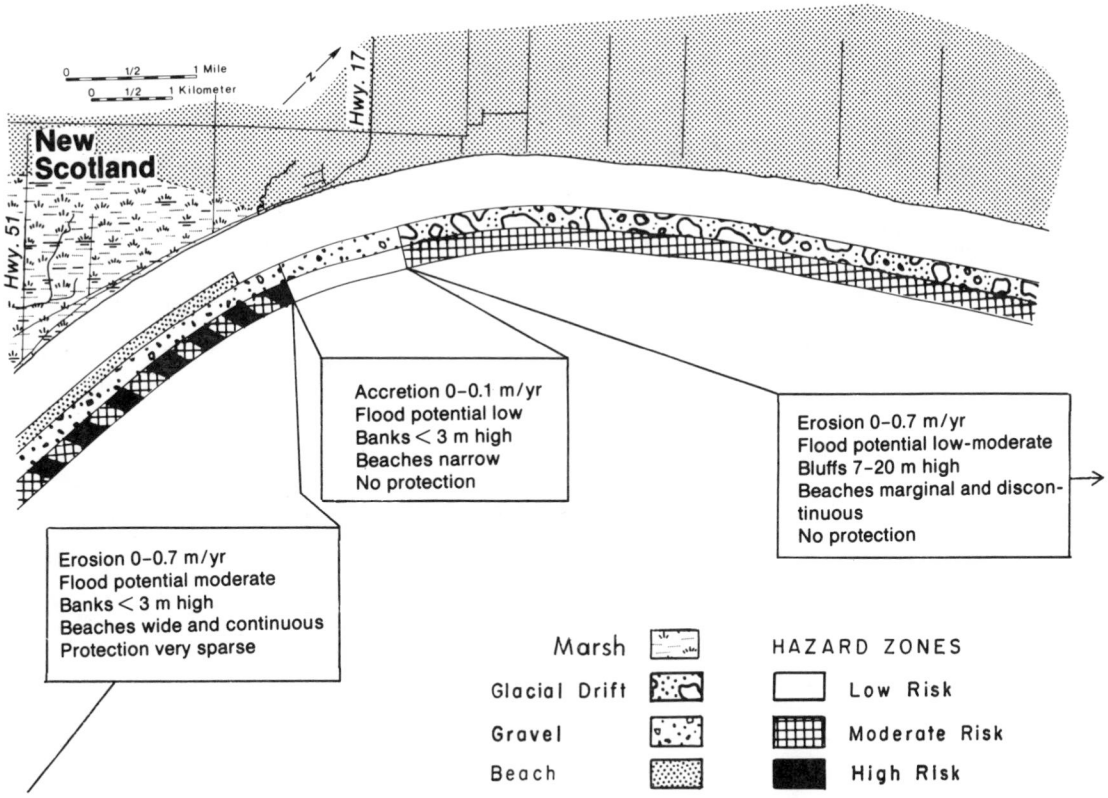

New
Scotland

Hwy. 51

Hwy. 17

0   1/2   1 Mile
0   1/2   1 Kilometer

Accretion 0–0.1 m/yr
Flood potential low
Banks < 3 m high
Beaches narrow
No protection

Erosion 0–0.7 m/yr
Flood potential low-moderate
Bluffs 7–20 m high
Beaches marginal and discon-
tinuous
No protection

Erosion 0–0.7 m/yr
Flood potential moderate
Banks < 3 m high
Beaches wide and continuous
Protection very sparse

Marsh

Glacial Drift

Gravel

Beach

HAZARD ZONES

Low Risk

Moderate Risk

High Risk

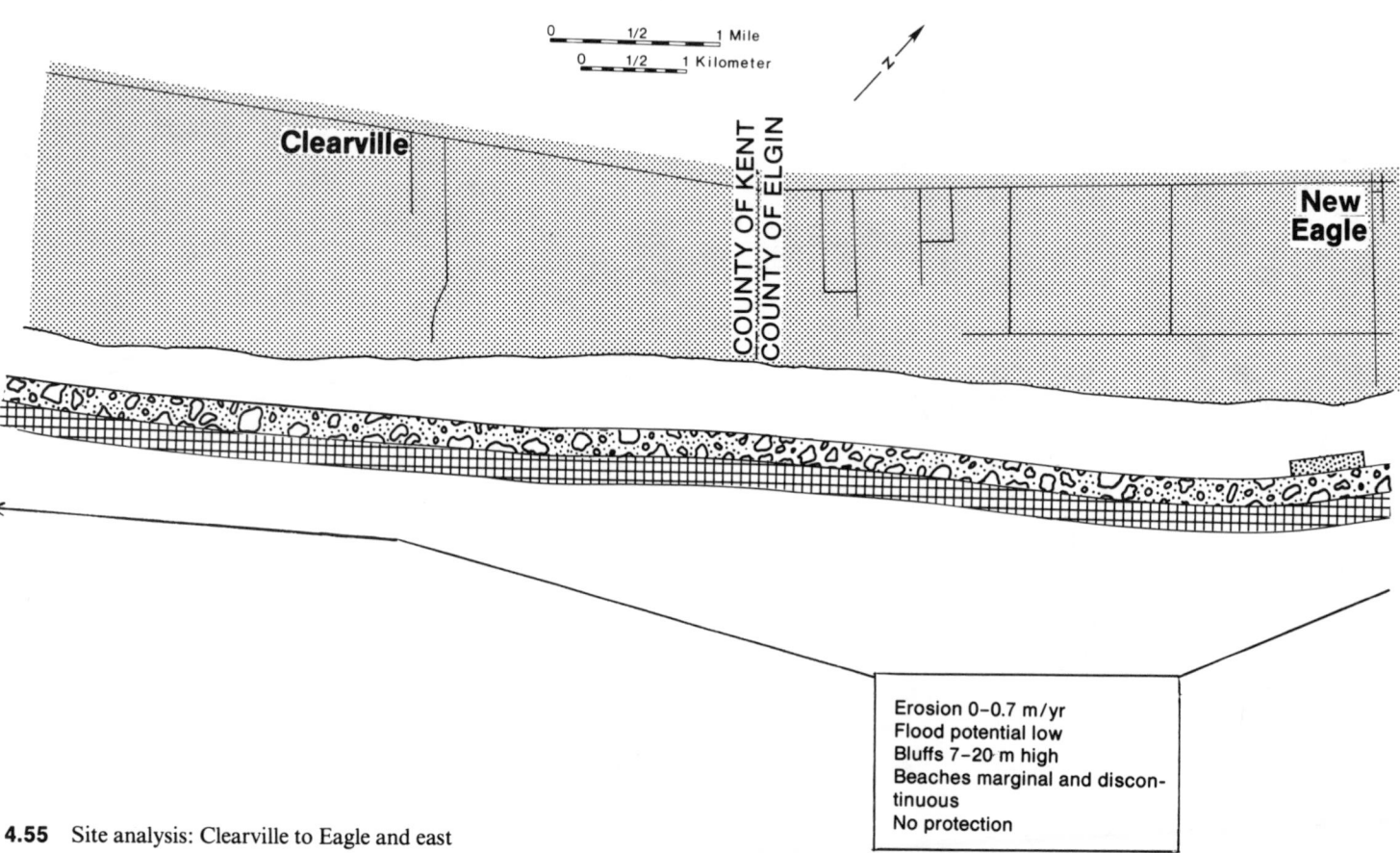

4.55 Site analysis: Clearville to Eagle and east

Clearville

New Eagle

COUNTY OF KENT
COUNTY OF ELGIN

0   1/2   1 Mile

0   1/2   1 Kilometer

N

Erosion 0–0.7 m/yr
Flood potential low
Bluffs 7–20 m high
Beaches marginal and discontinuous
No protection

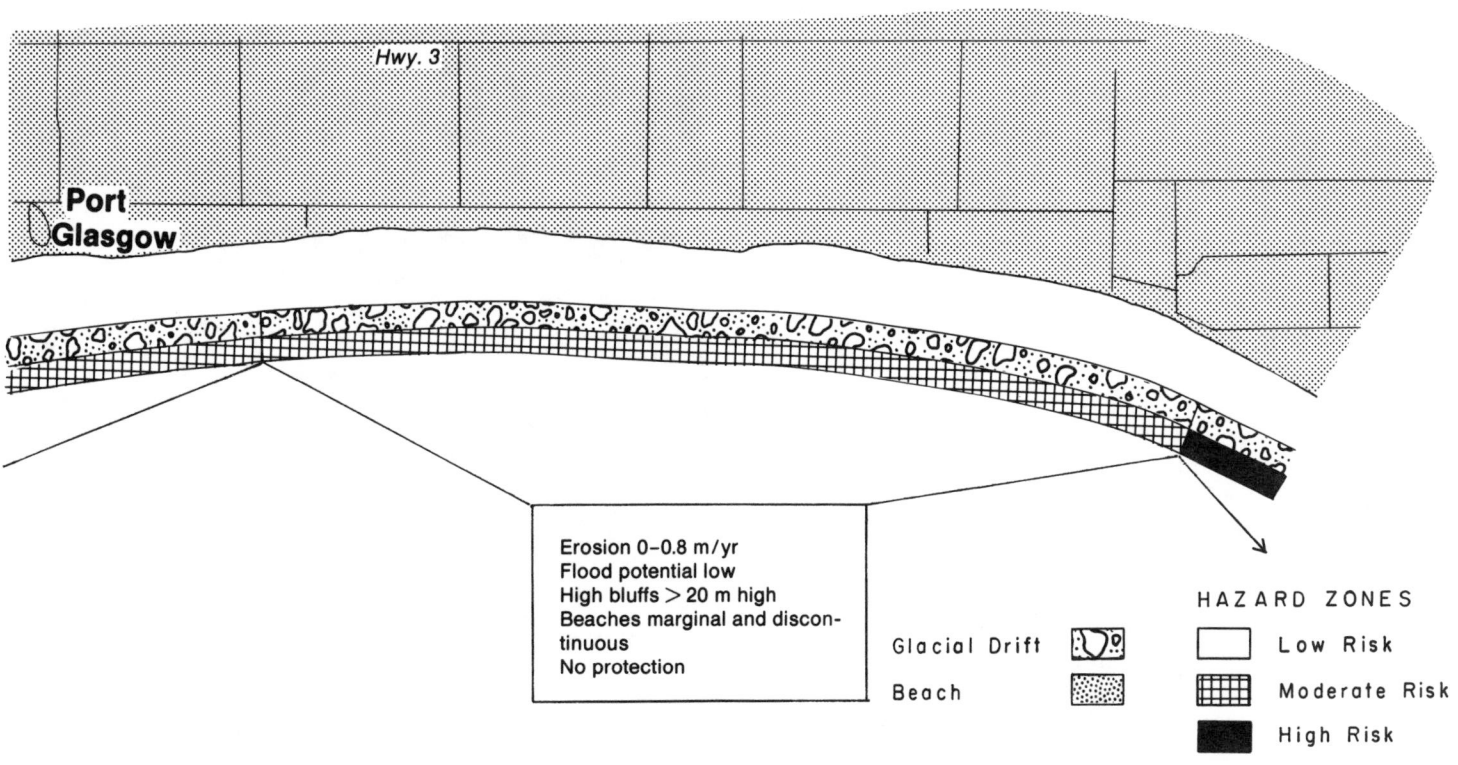

Erosion 0–0.8 m/yr
Flood potential low
High bluffs > 20 m high
Beaches marginal and discontinuous
No protection

Glacial Drift

Beach

HAZARD ZONES

Low Risk

Moderate Risk

High Risk

Reach 5 (Canada): County of Kent, Ontario

The length of shoreline in this county is about 90 kilometers, extending from Wheatley Harbour on the west almost to Port Glasgow on the east, including Pointe aux Pins (figs. 4.46 and 4.53 to 4.55). The shoreline is nearly linear with the exception of the sand spit protuberance of Pointe aux Pins at Rondeau Provincial Park.

Onshore relief is characterized by a gradual rise from low sand beaches at Wheatley to 20-meter-high bluffs at Port Alma. These bluffs are composed of till overlain by glaciolacustrine deposits, reaching a maximum height of 25 meters near Port Crewe, then gradually decreasing back to beach level at Erieau-Pointe aux Pins-Rondeau Provincial Park. East of Rondeau the bluff line gradually reemerges and continues again, in a nearly straight line, to the eastern limit of the county, rising to a height of just over 20 meters (fig. 4.56).

The physical processes that generated the landform of Pointe aux Pins are the mirror images of the physical processes that generated the landform Point Pelee, a direct consequence of the bifurcated nearshore current observable near Port Crewe (reference 57, appendix B). Consistent with such mirror-image mechanisms are the well-documented relict beach ridges making up the western side of Point Pelee and its relict beach-ridge counterparts making up the eastern side of Pointe aux Pins (fig. 4.57).

Consequently, the formation of beaches between Point Pelee and Port Crewe are consistent with predominant nearshore cur-

4.56 Steep unvegetated bluff that is characteristic of the area east of Rondeau.
4.57 Infrared vertical aerial photograph of Rondeau illustrating the convergence of the relict beach ridges.

4.58 Reclaimed agriculture land with extensive cottage development west of Rondeau Harbour, which can be seen to the top left of the photo.

rents from east to west; between Port Crewe and Erieau-Pointe aux Pins, with predominant nearshore currents from west to east; and east of Pointe aux Pins, with predominant nearshore currents from east to west.

The central portion of the reach between Pelee and Rondeau is somewhat unique to the Lake Erie shoreline in that it contains no major drainage creeks into the lake proper. This results in a long and narrow stretch of about 35 kilometers of quite uniform bluff terrain centered roughly on Port Crewe, the point of origin of the bifurcated nearshore current.

The portions of the shoreline of Kent County most vulnerable to erosion are the bluff areas at Port Alma (erosion rates as high as two meters per year). Other bluff areas such as those in the vicinity of Port Crewe, Ouvry, and Port Glasgow, have erosion rates limited to between 0.2 to 0.7 meters per year. The portions of the shoreline most vulnerable to flooding are the beach areas at Erieau and just east of Wheatley at the confluence of two creeks. Also susceptible to flooding is the total inner shoreline of Rondeau Harbour.

The beach area at and immediately west of Erieau is extensively protected from erosion by fields of groins of variable design and construction materials. This beach area has a high concentration of summer cottages, immediately behind which are some 650 hectares of marsh which have been drained and reclaimed for agricultural purposes since 1913. A dike road separates the farming region from the beach and cottage region (fig. 4.58). Thus the agricultural areas are afforded protection from flooding and the recreational areas are excluded from such protection. This region near Erieau is, however, the only region in Kent County where intensive agricultural uses occur on depression lands. The rest of the agricultural activities that occupy the Kent County shoreline take place on bluff regions. Very few shoreline protective structures exist in these areas.

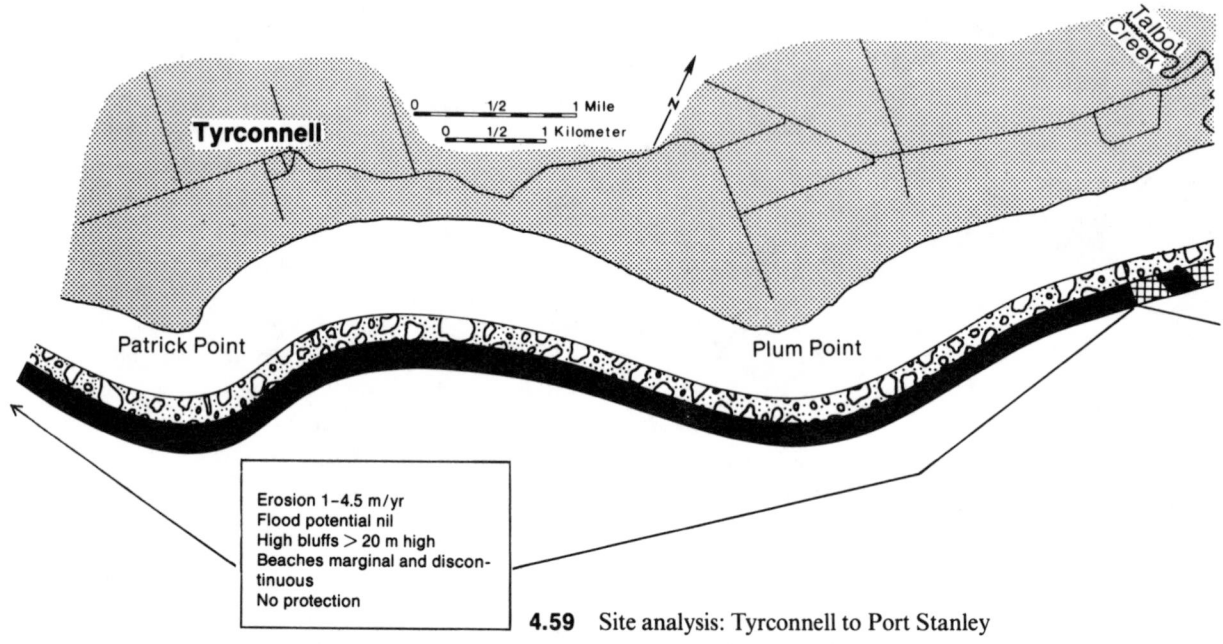

Erosion 1–4.5 m/yr
Flood potential nil
High bluffs > 20 m high
Beaches marginal and discontinuous
No protection

4.59   Site analysis: Tyrconnell to Port Stanley

Reach 6 (Canada): County of Elgin, Ontario

The length of the shoreline in this county is approximately 90 kilometers, extending from Port Glasgow on the west to just beyond Port Burwell on the east (figs. 4.55 and 4.59 to 4.61).

This distinctively concave region is almost exclusively characterized by bluffs of the highest and steepest slopes to be found

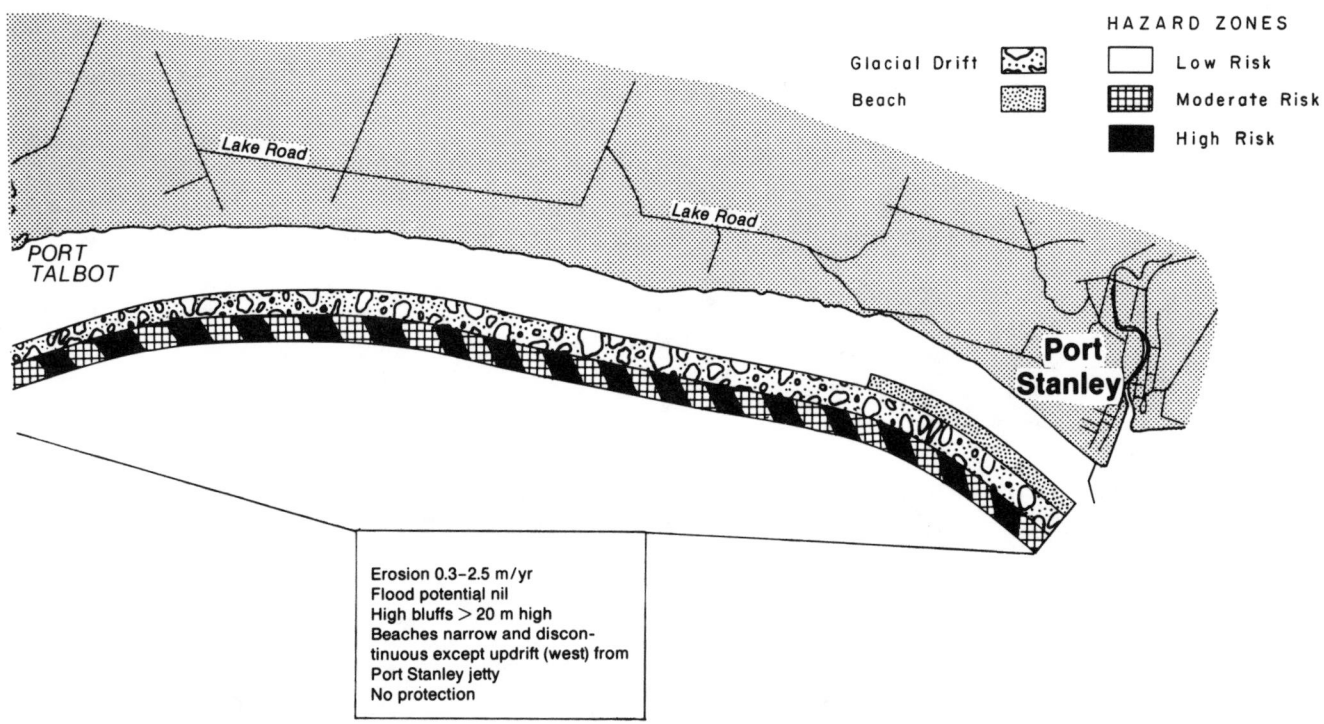

HAZARD ZONES

Glacial Drift

Beach

Low Risk

Moderate Risk

High Risk

Lake Road

Lake Road

PORT
TALBOT

**Port
Stanley**

Erosion 0.3–2.5 m/yr
Flood potential nil
High bluffs > 20 m high
Beaches narrow and discon-
tinuous except updrift (west) from
Port Stanley jetty
No protection

along the Canadian shore of Lake Erie (fig. 4.62). They vary from over 20 meters high west of Port Glasgow to Patrick Point, decreasing to 10 meters at Plum Point, and rising up again to near 40 meters just west of Port Stanley. The bluffs continue at this height to Port Bruce, slightly decreasing to 26 meters at Port Burwell and continuing at this elevation to the eastern

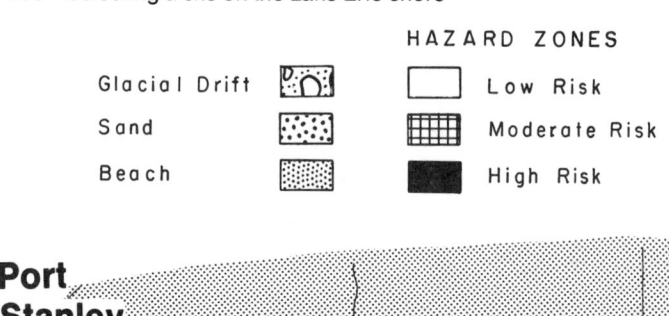

HAZARD ZONES

Glacial Drift   Low Risk

Sand   Moderate Risk

Beach   High Risk

0   1/2   1 Mile

0   1/2   1 Kilometer

N

Port Stanley

Dexter

Barnum Gulley

Erosion 0.2–0.3 m/yr
Flood potential nil
Bluffs 7–20 m high
Beach marginal and discontinuous
No protection

4.60   Site analysis: Port Stanley to Lakeview

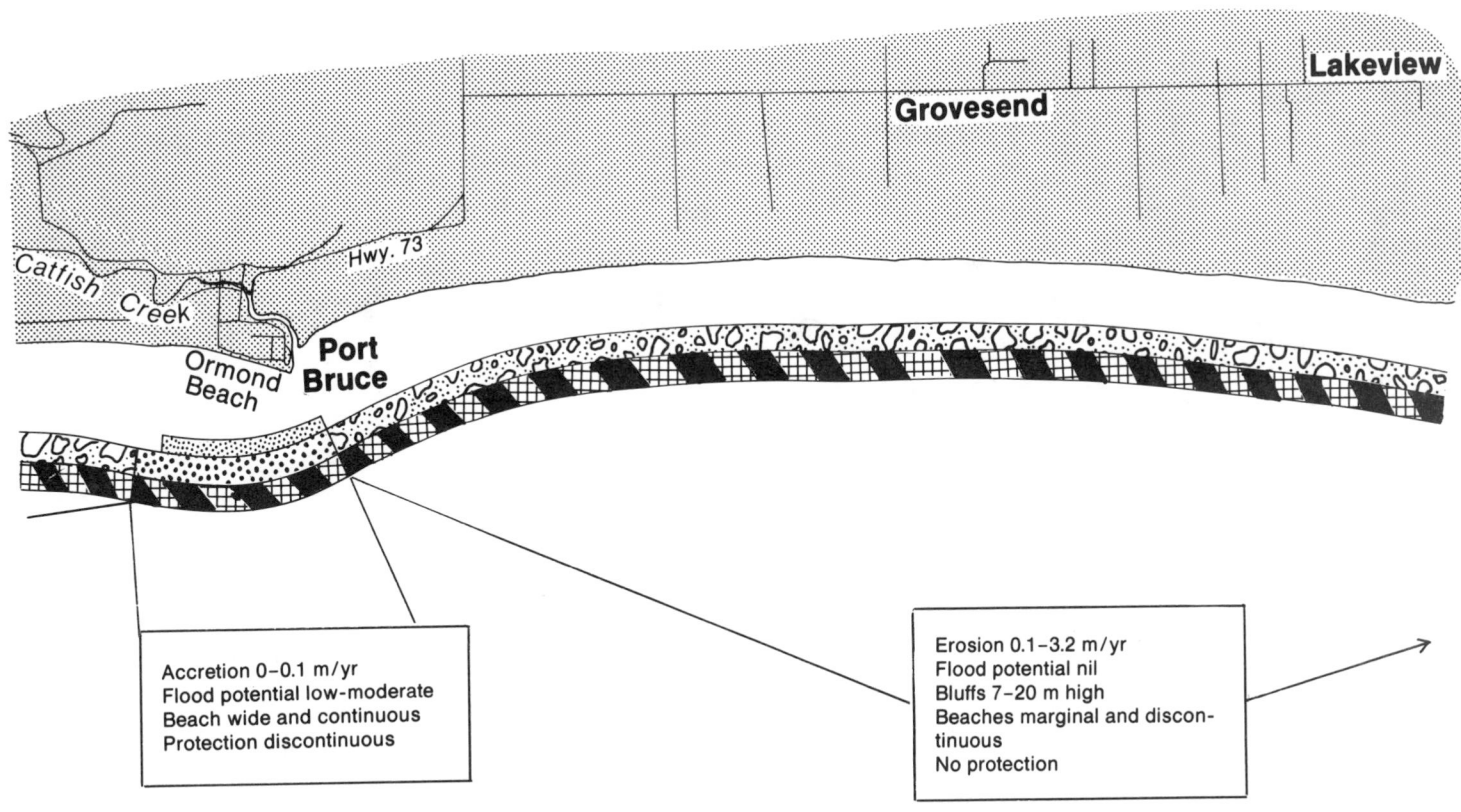

Accretion 0–0.1 m/yr
Flood potential low-moderate
Beach wide and continuous
Protection discontinuous

Erosion 0.1–3.2 m/yr
Flood potential nil
Bluffs 7–20 m high
Beaches marginal and discon-
tinuous
No protection

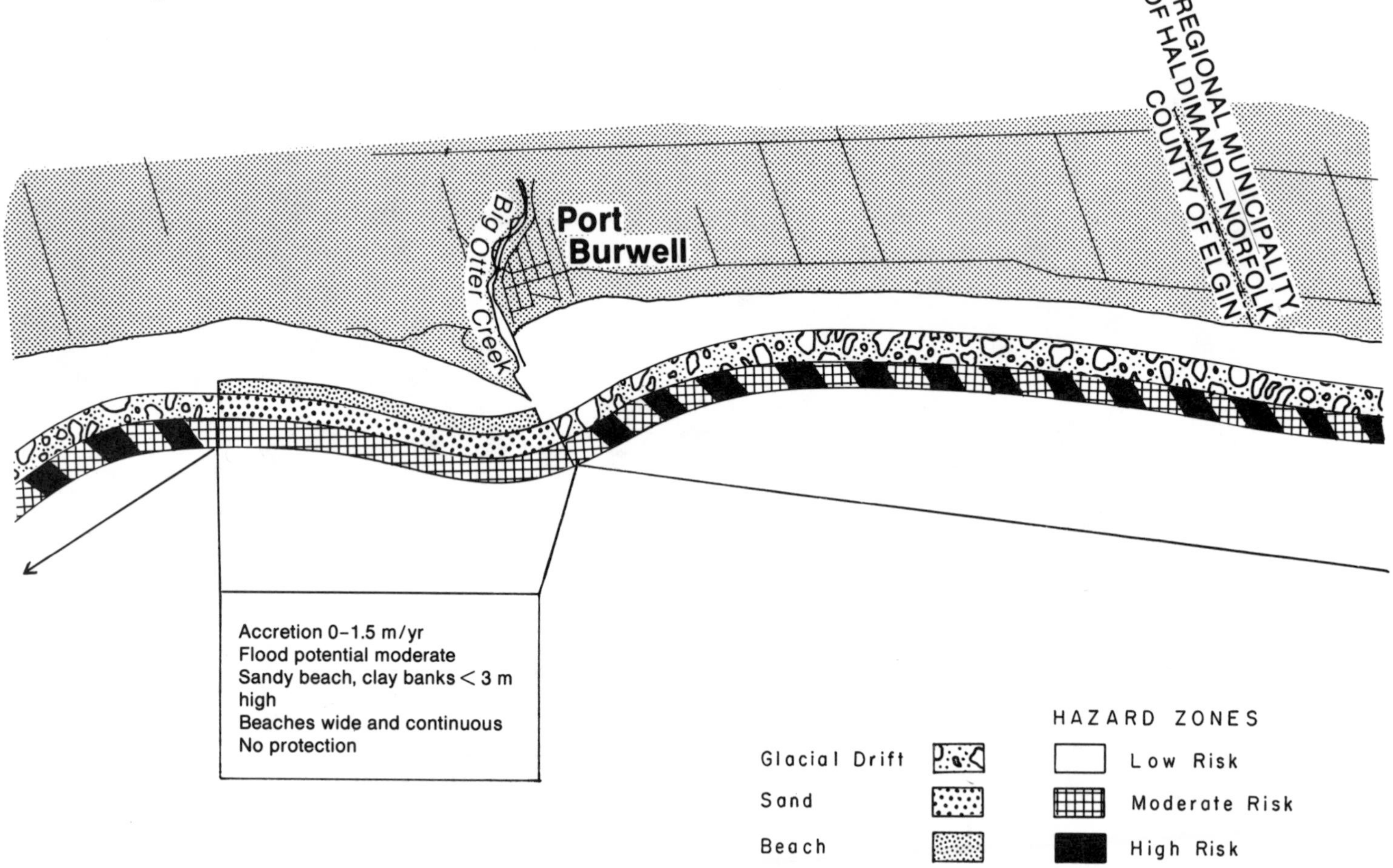

REGIONAL MUNICIPALITY
OF HALDIMAND—NORFOLK
COUNTY OF ELGIN

Big Otter Creek

**Port
Burwell**

Accretion 0–1.5 m/yr
Flood potential moderate
Sandy beach, clay banks < 3 m
high
Beaches wide and continuous
No protection

HAZARD ZONES

Glacial Drift

Sand

Beach

Low Risk

Moderate Risk

High Risk

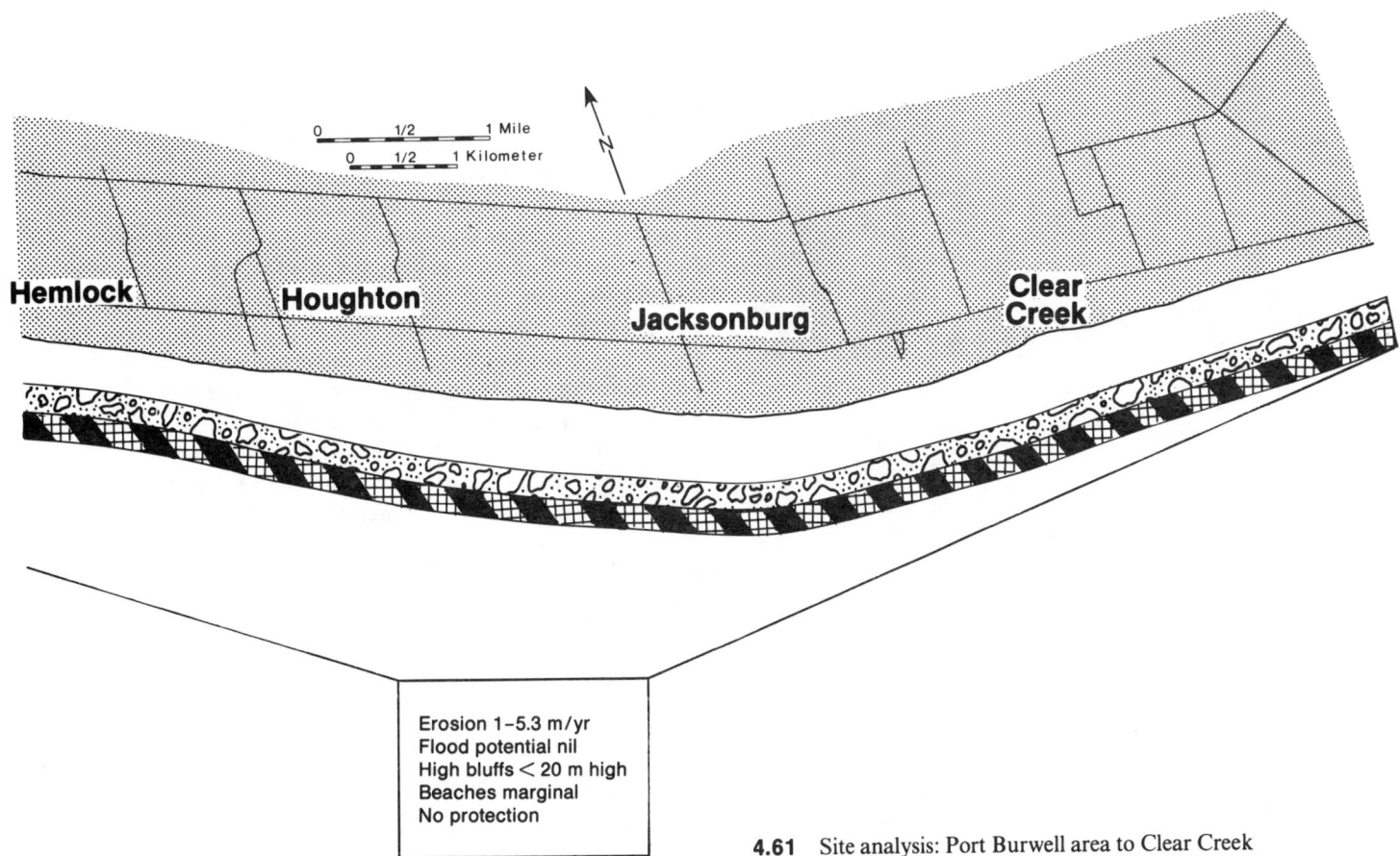

Hemlock

Houghton

Jacksonburg

Clear Creek

0   1/2   1 Mile

0   1/2   1 Kilometer

N

Erosion 1–5.3 m/yr
Flood potential nil
High bluffs < 20 m high
Beaches marginal
No protection

4.61   Site analysis: Port Burwell area to Clear Creek

boundary of the county. The sporadic, marginal, and discontinuous beaches provide virtually no protection from wave attack on the toe of the bluff. These bluffs, which are made up of till overlain by lacustrine silt and sand, often with an additional layer of wind-blown sand, are therefore subjected to the dynamic forces (waves, nearshore currents, sediment transport) that have created this dramatic concavity. Consequently, this stretch of Lake Erie shoreline is highly vulnerable to continuous erosion. The erosion rates, which range from 0 to 0.7 meters per year at Patrick Point to over 5 meters per year east of Port Burwell, are the highest and most persistent erosion rates along the entire north shore of Lake Erie. Erosion processes and the transport of erodible materials are largely responsible for the generation and continuous evolution of the sand spit formations of Pointe aux Pins in the neighboring western county and Long Point in the neighboring eastern regional municipality. Further evidence of notable sediment transport is found in the volume of sand accumulated on the updrift (western) side of the entrance piers of the three harbors: Port Stanley, Port Bruce, and Port Burwell (fig. 4.63).

4.62 Classic example of the physical characteristics of the central basin bluffs.

4.63 Accumulation of sand on the updrift side of Port Burwell pier.

These erosion mechanisms coupled with groundwater move-ent and the bluff's soil composition also result in several large gullies between Port Stanley and Port Burwell (i.e., Barnum Gully) (fig. 4.64).

The height and physical nature of the shore, combined with limited access to the lake, has prevented recreational development of this portion of shoreline. Thus land use remains predominantly agricultural. The farmlands are sufficiently elevated to preclude flood risks, despite their high erosion risk. To combat the erosion problem a continuous reinforcement of the base of the bluffs would be required. High costs discourage the establishment of such a continuous shoreline reinforcement project. The limited number of protective structures along this reach of shoreline attests to this dilemma.

Reach 7 (Canada): Regional Municipality of Haldimand-Norfolk, Ontario

This area is the longest of the shoreline counties, extending approximately 220 kilometers from just east of Port Burwell to just east of Mohawk Point, and includes possibly the most dramatic of the three distinctive Lake Erie's north shore landforms, Long Point. Long Point represents a natural demarcation between the intermediate-depth central basin of Lake Erie and the considerably deeper eastern basin. The Long Point landform extends 40 kilometers eastward into the deepest portion of the lake (figs. 4.61 and 4.65 to 4.68). Much of the area is parkland

4.64 Gullies can develop to the extent of acquiring a name, such as the Barnum Gully.

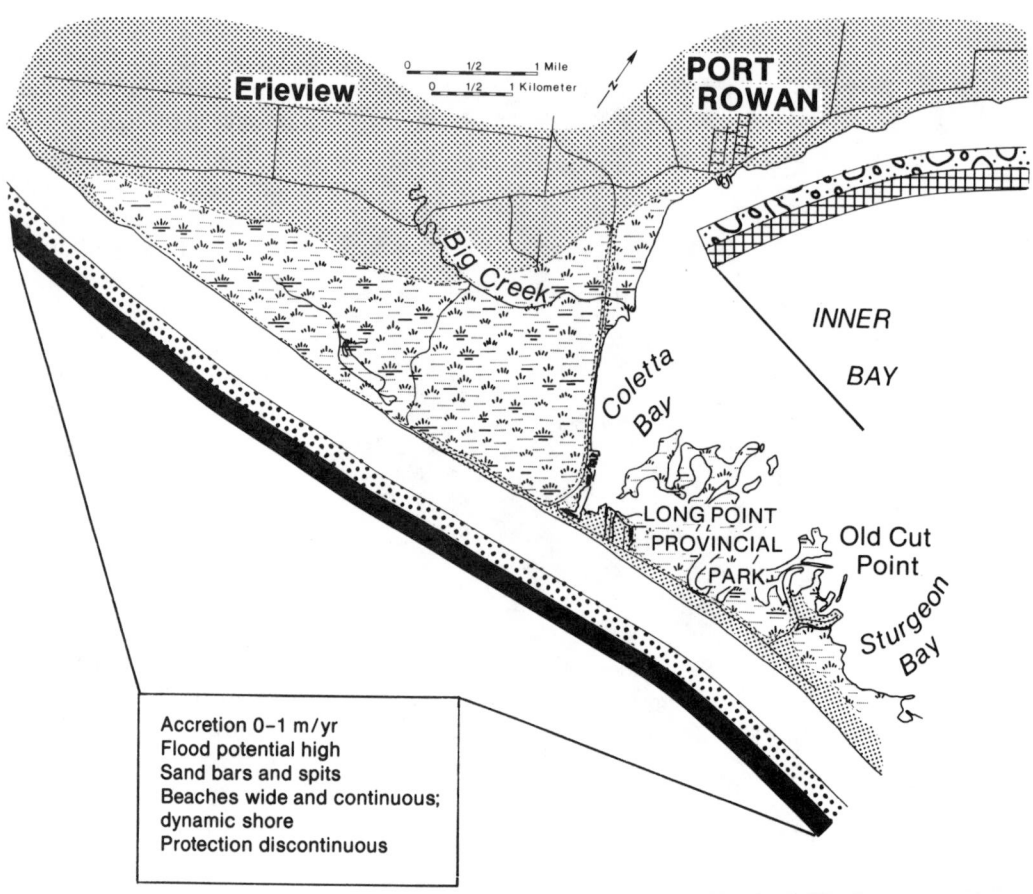

Accretion 0–1 m/yr
Flood potential high
Sand bars and spits
Beaches wide and continuous;
dynamic shore
Protection discontinuous

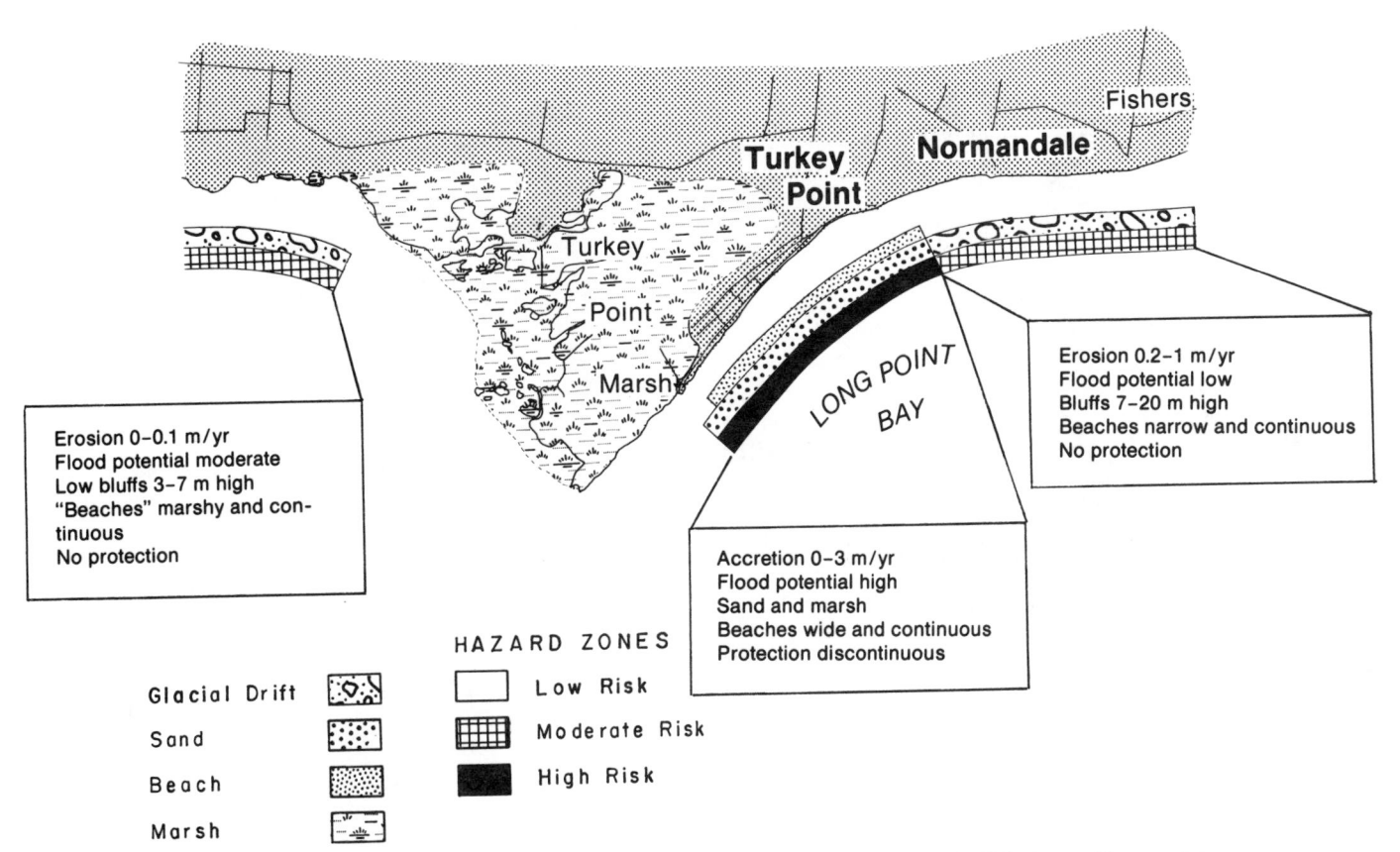

Erosion 0-0.1 m/yr
Flood potential moderate
Low bluffs 3-7 m high
"Beaches" marshy and con-
tinuous
No protection

Accretion 0-3 m/yr
Flood potential high
Sand and marsh
Beaches wide and continuous
Protection discontinuous

Erosion 0.2-1 m/yr
Flood potential low
Bluffs 7-20 m high
Beaches narrow and continuous
No protection

Glacial Drift
Sand
Beach
Marsh

HAZARD ZONES
Low Risk
Moderate Risk
High Risk

4.65 Site analysis: Erieview to Normandale

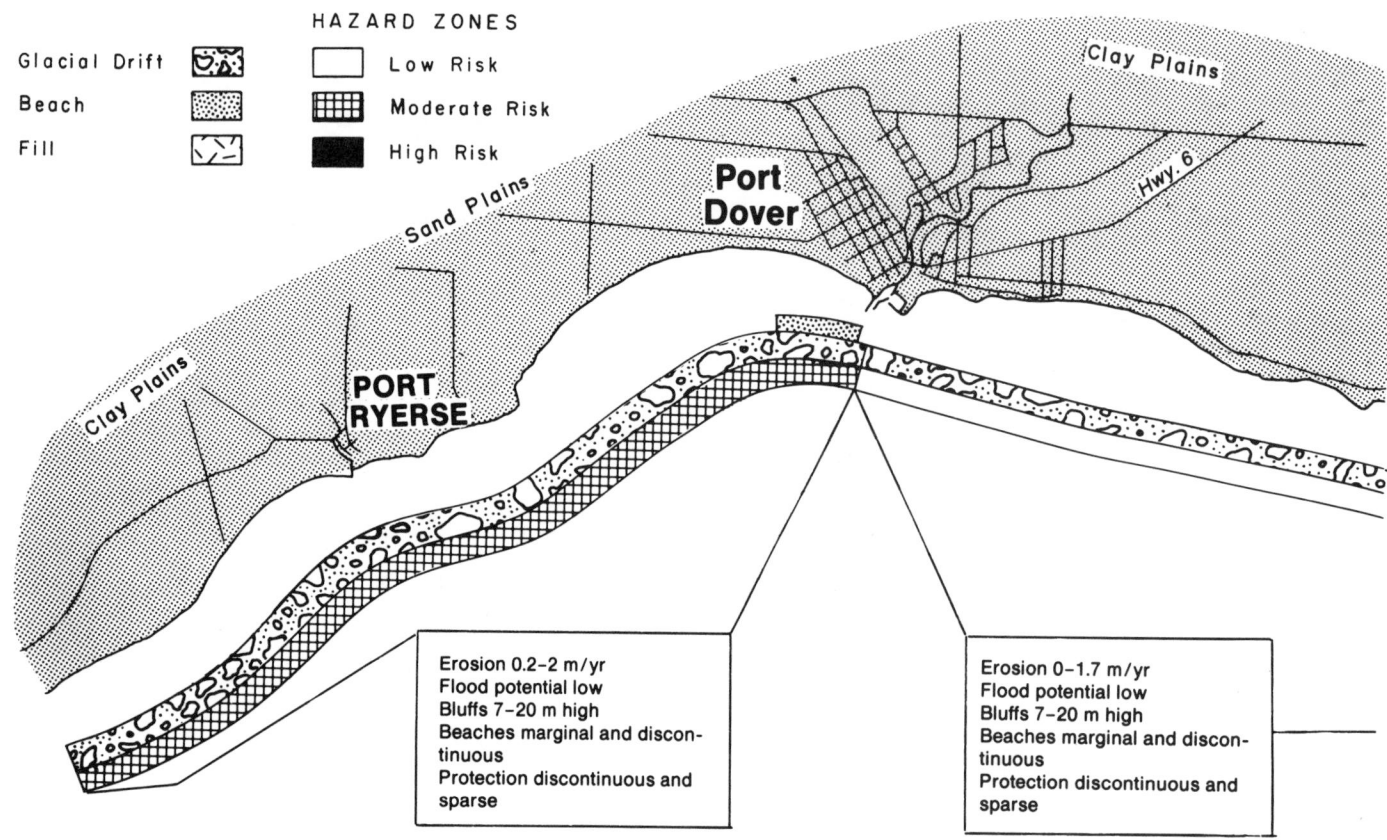

HAZARD ZONES

Glacial Drift

Beach

Fill

Low Risk

Moderate Risk

High Risk

Clay Plains

Sand Plains

Port
Dover

Hwy. 6

Clay Plains

PORT
RYERSE

Erosion 0.2–2 m/yr
Flood potential low
Bluffs 7–20 m high
Beaches marginal and discon-
tinuous
Protection discontinuous and
sparse

Erosion 0–1.7 m/yr
Flood potential low
Bluffs 7–20 m high
Beaches marginal and discon-
tinuous
Protection discontinuous and
sparse

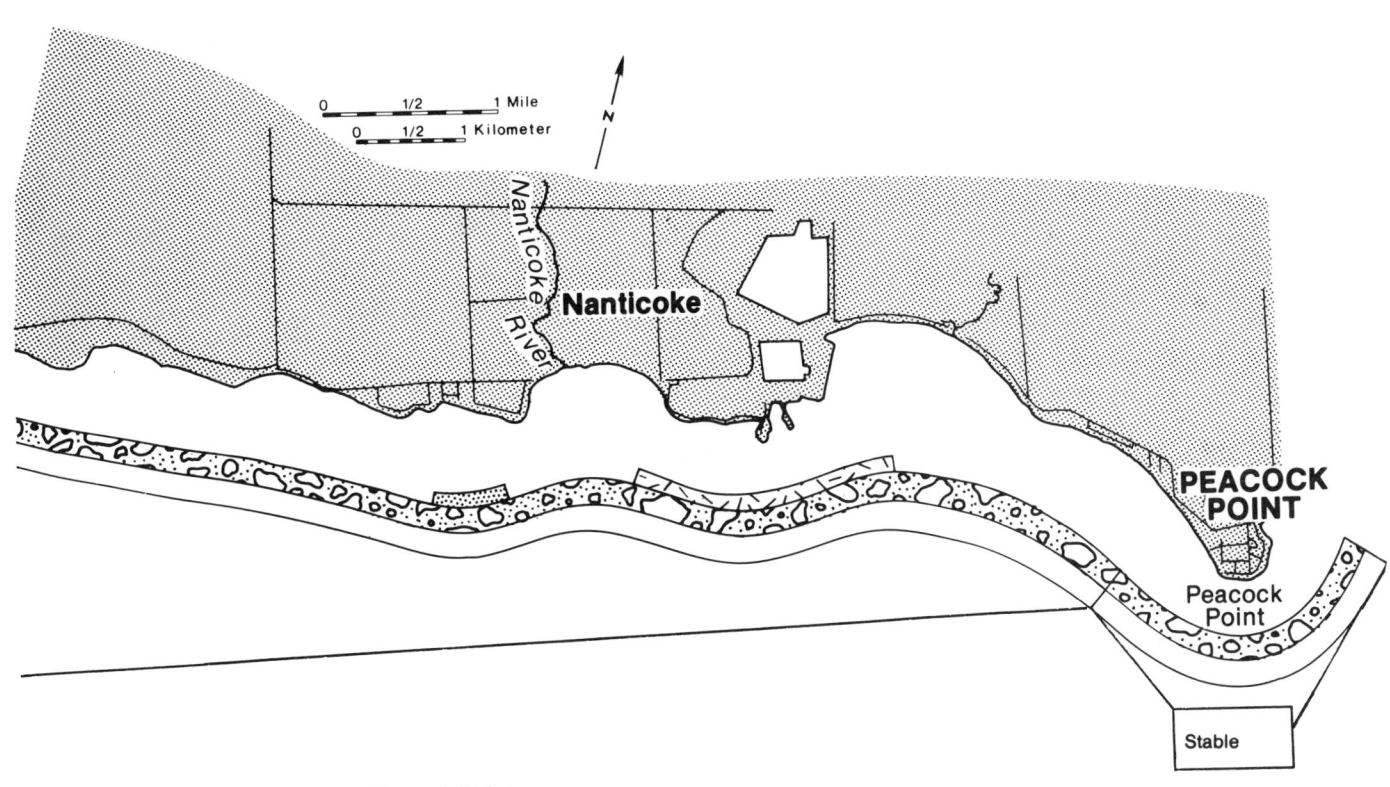

0   1/2   1 Mile

0   1/2   1 Kilometer

N

Nanticoke River

Nanticoke

PEACOCK POINT

Peacock Point

Stable

4.66 Site analysis: Port Ryerse to Peacock Point

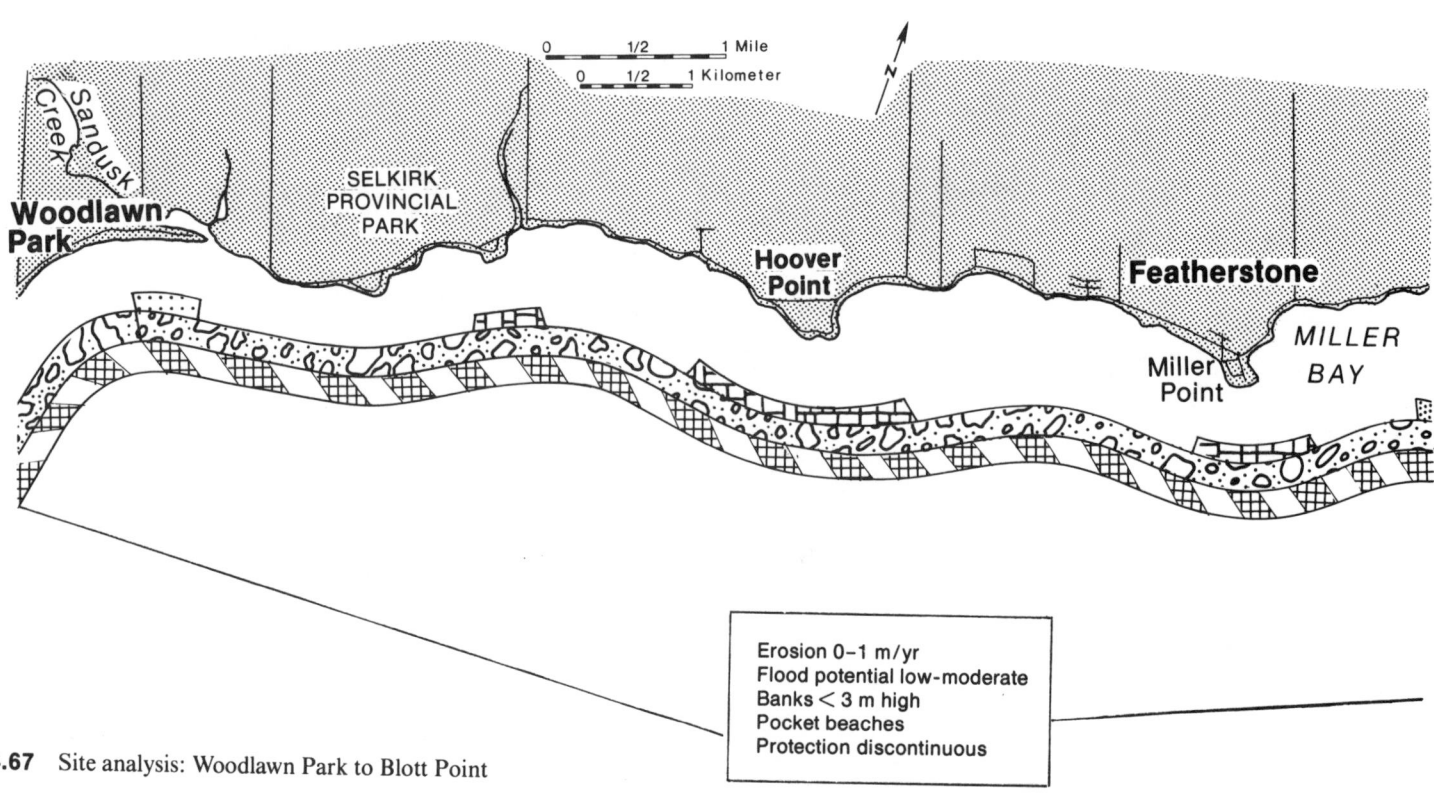

0   1/2   1 Mile

0   1/2   1 Kilometer

Sandusk Creek

Woodlawn Park

SELKIRK PROVINCIAL PARK

Hoover Point

Featherstone

Miller Point

MILLER BAY

Erosion 0–1 m/yr
Flood potential low-moderate
Banks < 3 m high
Pocket beaches
Protection discontinuous

4.67 Site analysis: Woodlawn Park to Blott Point

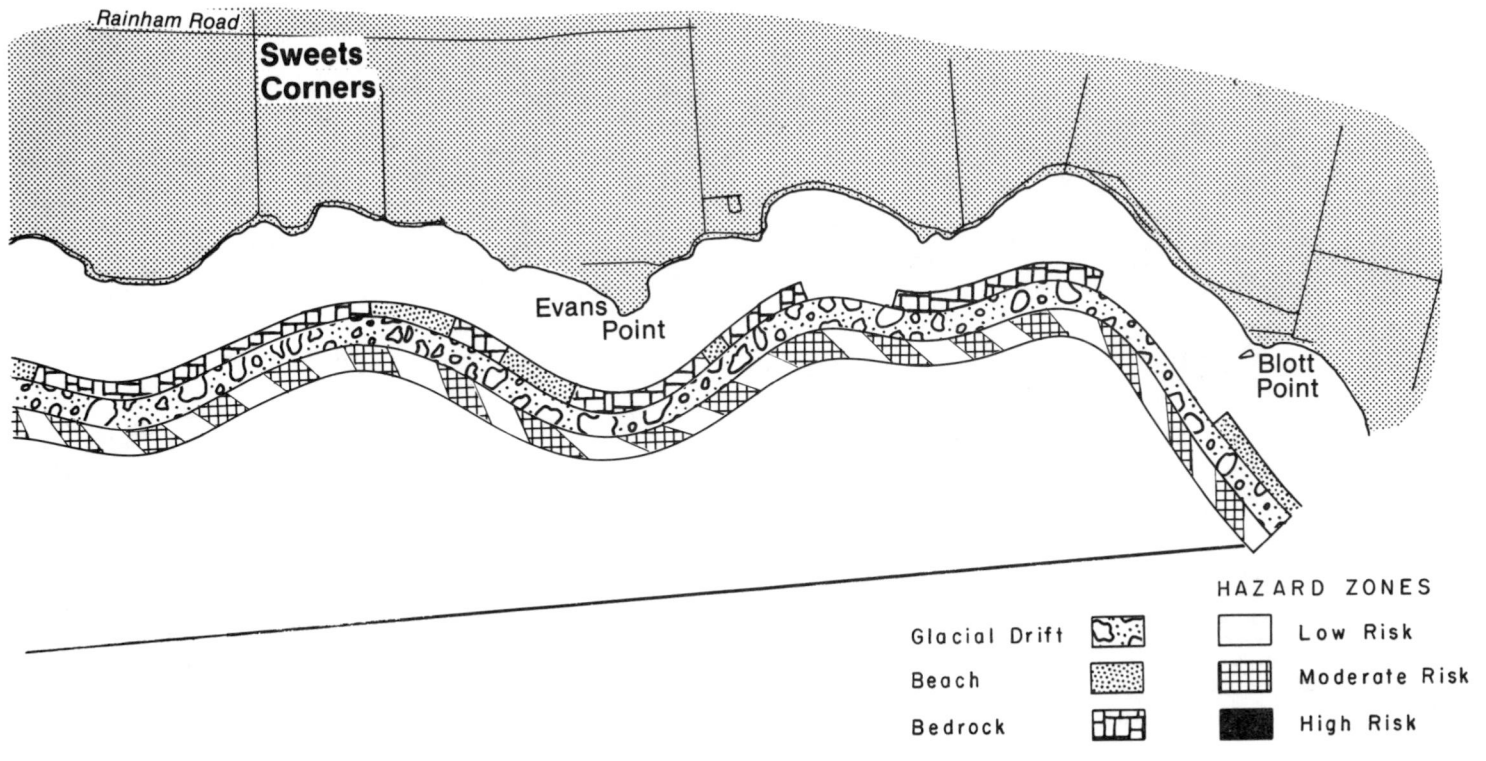

Rainham Road

Sweets Corners

Evans
Point

Blott
Point

HAZARD ZONES

Glacial Drift

Beach

Bedrock

Low Risk

Moderate Risk

High Risk

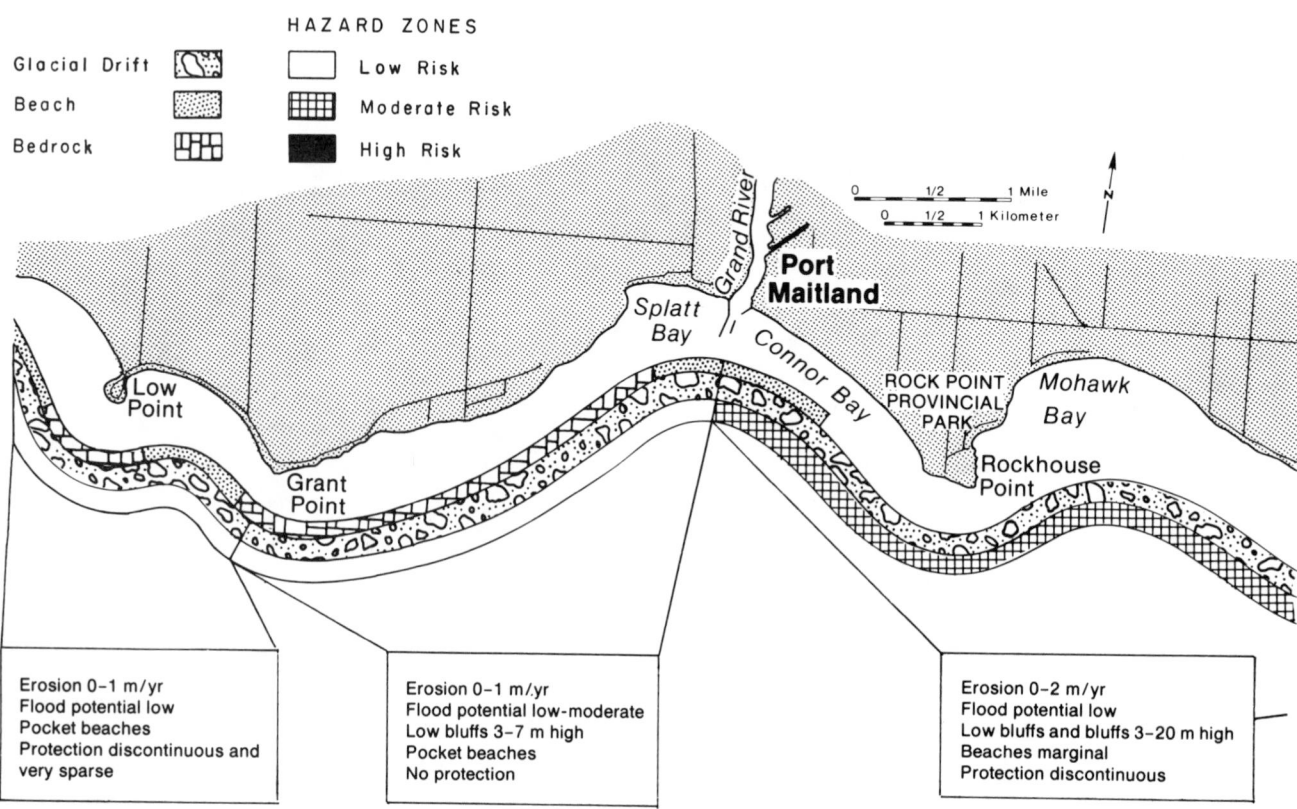

HAZARD ZONES

Glacial Drift   Low Risk

Beach   Moderate Risk

Bedrock   High Risk

Low Point

Grant Point

Splatt Bay

Grand River

Port Maitland

Connor Bay

ROCK POINT PROVINCIAL PARK

Mohawk Bay

Rockhouse Point

0   1/2   1 Mile

0   1/2   1 Kilometer

N

Erosion 0–1 m/yr
Flood potential low
Pocket beaches
Protection discontinuous and
very sparse

Erosion 0–1 m/yr
Flood potential low-moderate
Low bluffs 3–7 m high
Pocket beaches
No protection

Erosion 0–2 m/yr
Flood potential low
Low bluffs and bluffs 3–20 m high
Beaches marginal
Protection discontinuous

4.68 Site analysis: Low Point to Burnaby

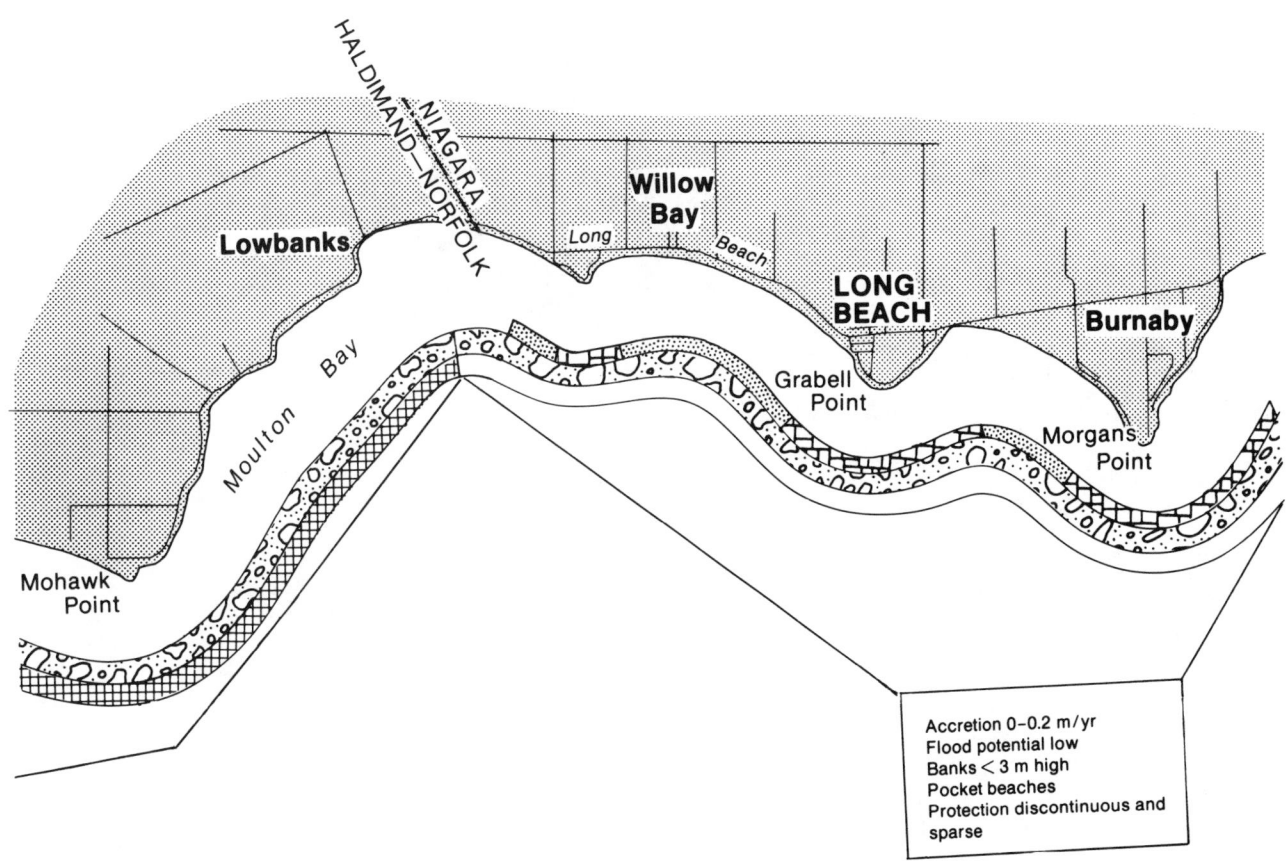

Accretion 0–0.2 m/yr
Flood potential low
Banks < 3 m high
Pocket beaches
Protection discontinuous and
sparse

4.69   Lacustrine sands constituting the entire bluff at Sand Hill Park mark the highest elevations on the entire north shore.

and a national wildlife preserve, not available for development and therefore not shown in the site analysis maps.

The relief of this region is like that of Elgin County to the immediate west. The steep bluffs continue eastward from Port Burwell and gradually increase in height from more than 20 meters to more than 55 meters at Sand Hill Park, the location with the highest elevation along the north shore of Lake Erie (fig. 4.69).

4.70   A typical view of clay bluffs at Clear Creek dropping to a height of 10 meters only 2 kilometers from Sand Hill.
4.71   Cottages occupying the beach ridge at the base of Long Point with Big Creek Marsh in the background.

As the height of the bluffs increases along the shoreline so also does the lacustrine sand content of the bluffs. This sand provides material for the prominent and unusual dunes rising high above the lake at Sand Hill Park. Eastward from Sand Hill Park, in a short span of only 8 kilometers, the height of the bluffs drops to 10 meters and their composition changes to almost all clay (fig. 4.70).

This abrupt change in shoreline configuration occurs at the western extremity or base of the Long Point landform, where littoral drift processes generate and continually reshape relict beach ridge patterns, which are the principal feature of Long Point sand spit relief. At this point simultaneous changes in shoreline orientation, shore composition, and height above lake level take place. Bluffs give way to lowlands; however, the bluff line does not disappear but turns inland and continues in an almost straight line towards Normandale, thus creating a back shore for the large area of more than 8000 hectares of marshland that Long Point sand spit encloses. These marshlands, which extend northward into Long Point Bay and Inner Bay, merge with the marsh area at the mouth of Big Creek. This marshy shoreline continues along Inner Bay to merge eventually with the marsh in the lee of Turkey Point, representing the longest continuous stretch of marshland to be found on the north shore (fig. 4.71).

Turkey Point, which is a much less imposing landform than Long Point, is characterized by a physical orientation roughly perpendicular to the orientation of Long Point. Whereas Long Point extends from west to east, Turkey Point follows the general curvature of Inner Bay and extends nearly from north to south in an apparent attempt to intersect the central portion of the Long Point spit formation at Pottohawk Point. The eventual build-up of Turkey Point would lead to the complete enclosure of Inner Bay and its associated marsh. A sandbar already extends across the bay making navigation for boats with deep draft difficult. East of the beach at Turkey Point, bluffs rise abruptly to heights in excess of 20 meters near Normandale. This abrupt reemergence of the bluffs is a direct consequence of the reconvergence of the Lake Erie shoreline and the bluffline of the Norfolk sand plain after they had temporarily diverged to accommodate the formation of the Long Point spit and its associated marshes. From Normandale the bluffs decrease in height to less than 10 meters at Fishers Glen, a distance of just over one kilometer, remaining close to that height to Port Ryerse, a distance of five kilometers. From here the bluffs gradually increase in height to 20 meters just west of Port Dover harbor. The beaches along this stretch of shoreline east of Normandale to Port Dover harbor are very narrow, but are generally wider east of Port Ryerse, where several small land projections assist both their development and stability. Only at Port Ryerse and Port Dover, where harbor jetties extend into the lake, are significant recreational beaches developed. At Port Dover, the Norfolk sand plain, which backs the shore to the west, gives way to the Haldimand clay plain, which is much lower in elevation near the lake shoreline than is the sand plain. This decrease in height is coincident with a slowly rising bedrock surface, which is just below lake level about five kilometers east of Port Dover, and outcrops to more than one meter above lake level within the next five kilometers. The bedrock is not continuous, but alternates with sand, gravel, or shingle beaches as far east as the Niagara River. These bedrock outcrops, which commence near Port Dover and extend along the eastern extent of the north shoreline of Lake Erie, are the controlling factors of shore relief in this region (figs. 4.72 and 4.73).

East of Port Dover the bluffs gradually decrease in height to about 10 meters at Peacock Point, beyond which they extend eastward at this height to Rockhouse Point. The irregular shore-line in this area is identified by rock outcrop headlands at Peacock, Hoover, Miller, Evans, Blott, Low, Grant, and Rockhouse Points. At Port Maitland, at the mouth of Grand River, the harbor jetties act like headlands, accumulating sandy beaches on both sides.

The remainder of the shoreline lying within this regional municipality includes Mohawk Bay and Moulton Bay, which are separated by Mohawk Point. Along Mohawk Bay the bluffs range in height from 10 to 17 meters. Near the bay-head shore, the road has been eroded, requiring that it be rebuilt further inland. East of Mohawk Point, along the Moulton Bay shoreline near Lowbanks, the bluff essentially disappears. The bluffs

4.72 A house under construction near Port Dover in spite of the visibly eroding bluff.

4.73 The rock outcrop at Evans Point is a typical example of the shoreline in this region.

change to low clay banks up to two meters high, usually fronted by gravelly, shingle, or bedrock shores. These beaches are washed over with each major southwesterly storm (figs. 4.74 and 4.75).

In general, considering the entire shore of Haldimand-Norfolk regional municipality, two dominant types of surface deposits over the bedrock are found: sand and clay. The Haldimand clay plains stretch from east of Port Dover to well beyond the eastern limit of this municipality. This clay plain also fronts the Norfolk sand plain at the western end (base of Long Point) and east of Turkey Point. Apart from the extensive reach of the Haldimand clay plain and the two comparatively small shoreline segments of the Norfolk sand plain, the remainder of the shoreline is controlled by numerous outcrops of limestone, which form erosion-resistant headlands. These headlands, which are found predominantly east of Port Dover, separate a series of bays, most of which contain sheltered, stable beaches. These beaches are not subject to erosion but are highly vulnerable to wave attack. Discontinuous shore protection in the form of individual groins and seawalls currently exists. Although they do not provide complete protection to this reach of shore, these groins and seawalls do afford some protection against wave uprush during southwesterly storms.

Most of this shoreline has attracted recreational activity and ensuing development. Despite the shoreline bluffs, erosion along this stretch of Lake Erie is generally not a serious problem, with long-term erosion rates averaging between 0.1 to 0.3 meters per year. Exceptions are found near Port Dover and Mohawk Bay,

4.74   A view of the continuously eroding bluffs near Mohawk Point, where residential development has persisted.

4.75   A typical cottage development at Lowbanks, where sand beaches have formed between rock outcrop headlands.

along the southwestern shore of Long Point along Hastings Drive, where cottages cling precariously to a narrow spit between Lake Erie and the Big Creek Marsh, and the northeastern shore by the lighthouse near the tip of Long Point. At Port Dover and Mohawk Bay the shore recedes about 2 meters per year, whereas on Long Point shore storm-induced erosion of 5 to 15 meters is not uncommon. In 1985 it was necessary to relocate the lighthouse keeper's house from Long Point to the mainland—this despite a field of concrete groins that was constructed to stabilize the shore in the immediate vicinity of the lighthouse and the keeper's house! The groins themselves did not withstand the onslaught of waves (fig. 4.76).

Long Point has a history of vulnerability and submissive response to catastrophic climatic disturbances. Two such recent catastrophic disturbances were the devastating storms of April 1985 and January 1986, which resulted in total destruction of some cottages and in extensive property damage. One of the cottages on Hastings Drive was literally uprooted and transported by wind and wave over marshland for a distance of two kilometers into Big Creek Marsh (fig. 4.77).

Roy Binkley, one of the cottage owners along Hastings Drive, recalls that more than 100 meters of beach fronted his property when he built in 1955. "My wife even complained we were too far from the water." Since then he has spent about CN$10,000 on erosion control measures, including construction of a CN$6,000 breakwall, and has moved the cottage back from the water 12 meters (*London Free Press*, July 1985). The Dominion Wreck

4.76   Site of Long Point's lighthouse and keeper's house prior to out-flanking of groins by storm waves.

4.77   By the time the Good Friday 1985 wind and wave storm sub-sided, one of the Hastings Drive cottages had a new address, Big Creek Marsh.

Register at Ottawa and ancient mariners of the area say that since 1679, when the French explorer LaSalle's *Griffon* almost sank, more than 200 ships have gone to the bottom near Long Point! As recently as April 30, 1984, during a storm with winds gusting to 112 kilometers per hour and waves as high as 3 meters,

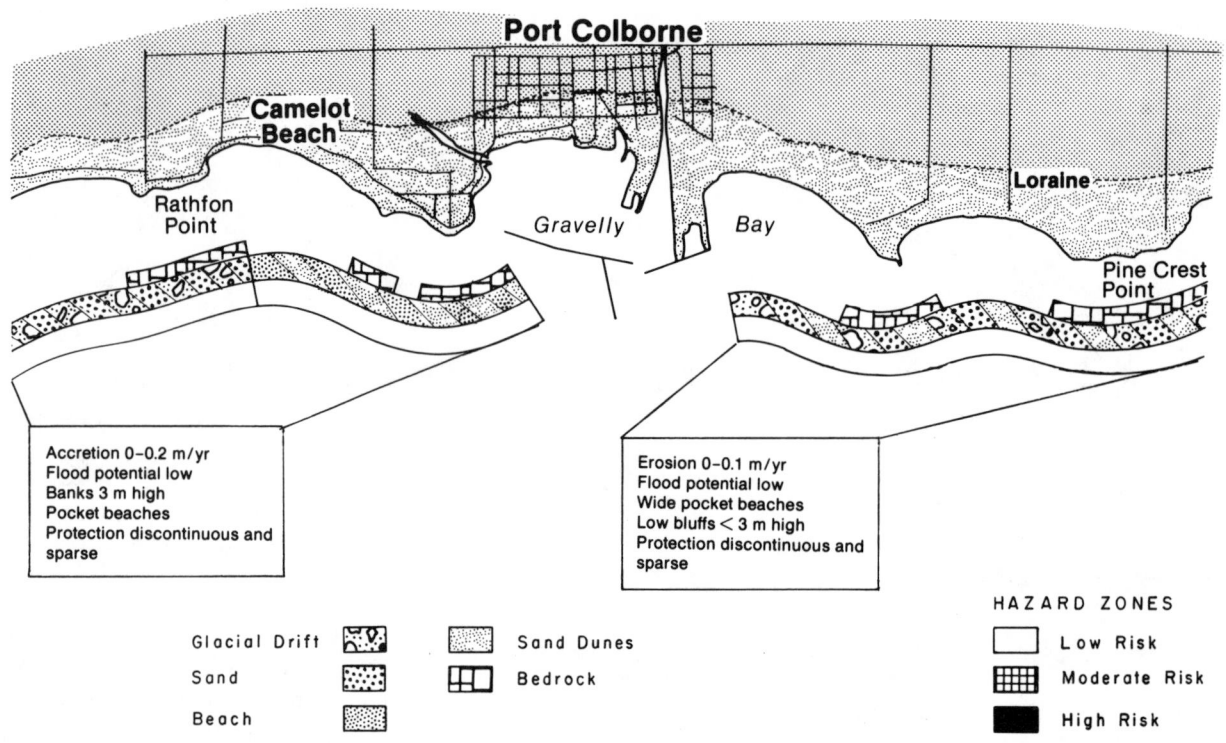

Port Colborne

Camelot Beach

Rathfon Point

Gravelly *Bay*

Loraine

Pine Crest Point

Accretion 0-0.2 m/yr
Flood potential low
Banks 3 m high
Pocket beaches
Protection discontinuous and sparse

Erosion 0-0.1 m/yr
Flood potential low
Wide pocket beaches
Low bluffs < 3 m high
Protection discontinuous and sparse

HAZARD ZONES

Glacial Drift

Sand

Beach

Sand Dunes

Bedrock

Low Risk

Moderate Risk

High Risk

4.78 Site analysis: Rathfon Point to Thunder Bay

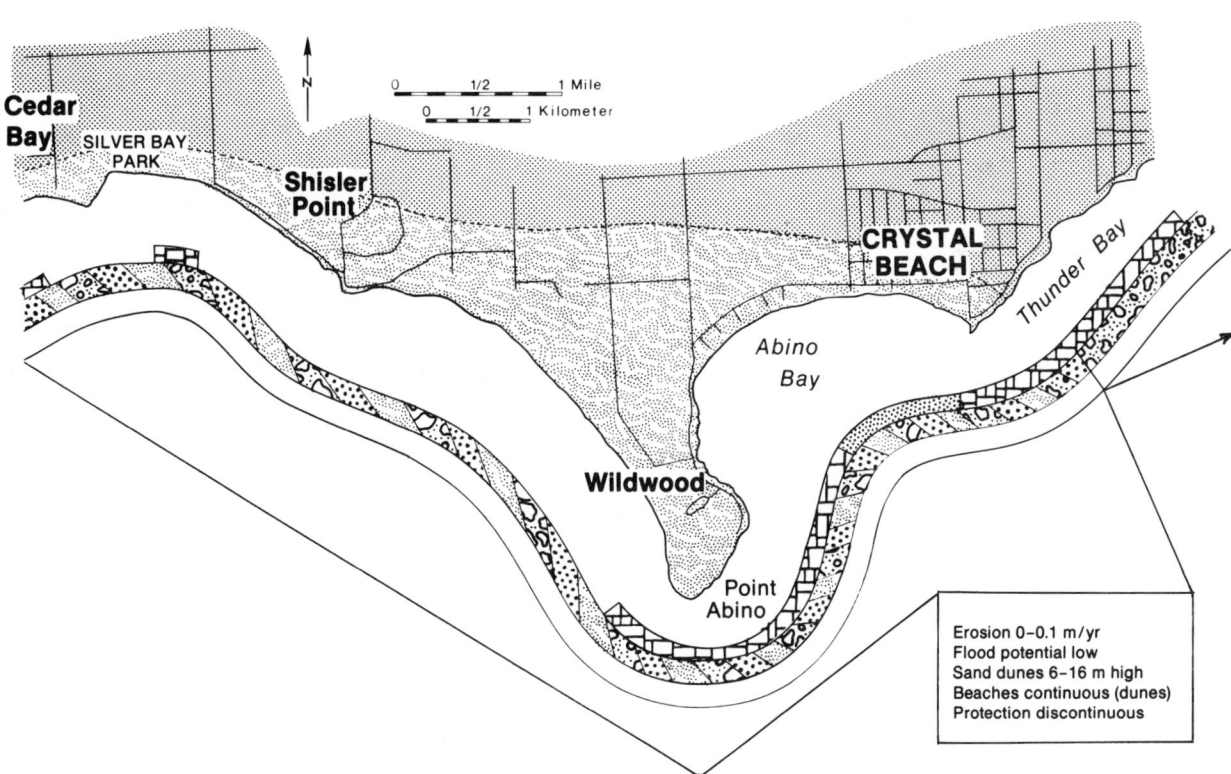

Cedar
Bay

SILVER BAY
PARK

Shisler
Point

CRYSTAL
BEACH

Thunder Bay

*Abino
Bay*

Wildwood

*Point
Abino*

0   1/2   1 Mile

0   1/2   1 Kilometer

N

Erosion 0-0.1 m/yr
Flood potential low
Sand dunes 6-16 m high
Beaches continuous (dunes)
Protection discontinuous

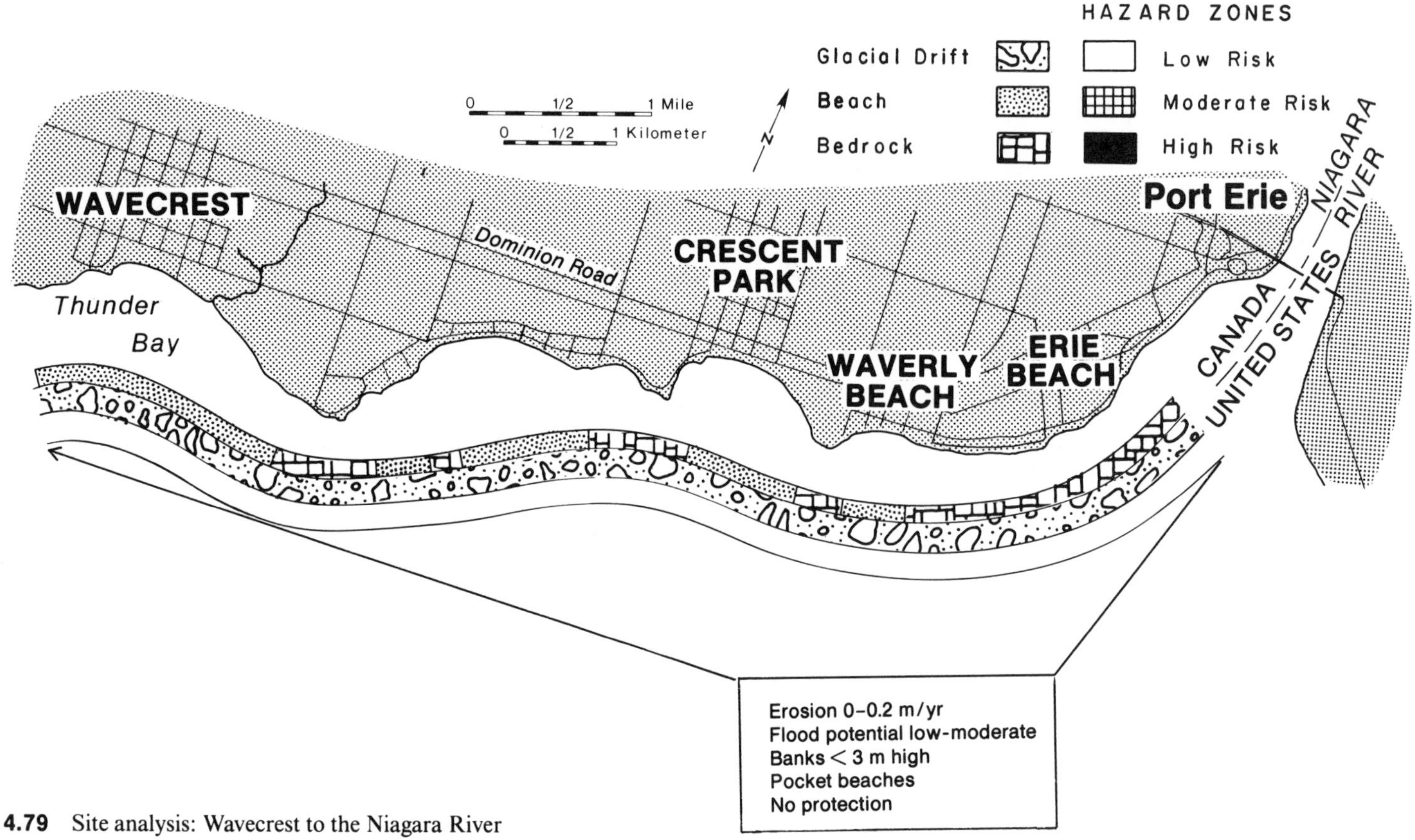

HAZARD ZONES

Glacial Drift · Low Risk

Beach · Moderate Risk

Bedrock · High Risk

0   1/2   1 Mile

0   1/2   1 Kilometer

N

WAVECREST

Dominion Road

CRESCENT PARK

Port Erie

NIAGARA RIVER

Thunder Bay

WAVERLY BEACH

ERIE BEACH

CANADA

UNITED STATES

Erosion 0–0.2 m/yr
Flood potential low-moderate
Banks < 3 m high
Pocket beaches
No protection

4.79   Site analysis: Wavecrest to the Niagara River

a commercial fishing boat, the 19-meter "Stanley Clipper," was capsized and sunk off Long Point, taking with it her crew of three.

Reach 8 (Canada): Regional Municipality of Niagara, Ontario

This regional municipality, the shortest of the north shoreline counties, extends approximately 60 kilometers from just east of Mohawk Point to the Niagara River (figs. 4.68, 4.78, and 4.79).

The irregular shoreline is accentuated by alternating rock outcrop headlands and indented sand or gravel beaches. In many parts of this eastern length of the north shore, the waters of Lake Erie have eroded the thin layer of glacial and lacustrine deposits that once covered the underlying bedrock, thus creating several headlands, including Grabell, Morgans, Rathfon, Sugar Loaf, Cassaday, Pine Crest, Shisler, and Windmill Points, among others. The largest headland is Point Abino, located between Shisler and Windmill Points, 15 kilometers east of the Welland Canal (fig. 4.80).

Stable sand or gravel beaches are found along the intervening bays and the headlands. The only departure from the general low relief of this stretch of shoreline is to the immediate east and west of Point Abino. Here sand dunes with heights upward to 20 meters are found. The bedrock underlying the eastern basin of Lake Erie renders the shoreline in this region much less susceptible to erosion than are its more westerly counterparts.

4.80 The bedrock outcropping near the water surface at Point Abino provides a permanent base for the lighthouse.

4.81 An example of some remedial measures used to protect the shoreline (Rathfon Point).

Erosion rates in this regional municipality are generally less than 0.2 meters per year, but there are some places in Wainfleet Township where the shore road has been washed out and rebuilt further inland, indicating that erosion can be a major concern. Flooding and wave uprush remain problematic, particularly during periods of high lake levels or storms. To protect against such ravages numerous seawalls and revetments or retaining embankments of various materials and designs have been constructed. These structures have achieved somewhat limited success due, in part, to inappropriate and ineffectual design and choice of construction materials (fig. 4.81).

Throughout this most eastern segment of the Canadian shoreline of Lake Erie summer residences and cottages abound. The entire shoreline is devoted to recreational activity. The attractiveness of the beaches and sand dunes, coupled with the majesty of the nearby Niagara Falls, has made this regional municipality a major center for tourists and vacationers (fig. 4.82).

Challenge of the shoreline

The dramatic configuration of the shoreline of Lake Erie is a direct consequence of the interactions of water and air (fig. 4.83). These physical processes often act in opposition to the farming, recreational, and commercial activities that support the human population clustered within the Lake Erie region. People, acting to protect their interests, construct protective structures, with variable success. Shoreline management depends not only upon

4.82   Good sandy beaches and the proximity to Niagara Falls and large urban centers create a need for recreational facilities such as Crystal Beach Amusement Park.

man's ability to properly construct such devices, but also upon man's ability to understand, combat, and harness the physical processes that prompt their construction. These tasks are not easy. They require insight, understanding, experimental and developmental manpower, money, climate and weather cooperation, political unity, and proper problem definition and agreement on methods and approaches to problem solving. The difficulty of meeting such requirements explains the variable success experienced by those tackling this problem.

One point is paramount: human actions factor into the dynamic equilibrium of the Lake Erie shore. Structures constructed to "protect" the shore can eventually cause more new problems than they ever solved. Prudent shore management requires a long view and respect for the people and processes involved.

4.83 A photo taken from Port Maitland during a storm onslaught (A) and calm (B) shows the extreme weather conditions to be expected on the shore of Lake Erie.

5 You and the shore: governmental programs and regulations

Human development of Lake Erie's coastal zone began in the early 1800s and accelerated in the early 1900s. During this time there was little, if any, regulation of activities along the lakeshore. However, increasing property losses, especially during the high lake levels of the 1950s and 1970s, as well as greater environmental consciousness with respect to conserving the coastal zone as a resource, has led to the enactment of several governmental programs and regulations.

Governmental agencies now regulate activities along the lakeshore in a number of ways and provide assistance to communities and property owners to reduce the impact of coastal hazards. In both the United States and Canada, policies and programs range from the federal to the local level. Assistance varies from forecasts of flooding by federal weather services to loan assistance for emergency house moving. The following pages provide a brief look at the most pertinent programs, regulations, and requirements for individual landowners. Additional information can be obtained by contacting the governmental agencies listed in appendix A.

Canada: federal programs

The Canadian federal government may cooperate with the province in the construction of flood and erosion control structures if conditions related to land-use zoning and costs and benefits are met. Cooperative projects include flood protection for agricultural land such as the dikes along Essex and Kent counties (Pelee, Mersea, and Harwich townships).

Flood damage assistance on a cost-sharing basis may be granted following a major disaster. Assistance includes the rebuilding of public structures such as highways and schools, and reimbursement to individuals for private losses, excluding losses covered by insurance or losses to seasonal homes. The federal government will also pay part of the cost of erosion protection if federal structures or commercial navigation in channels whose delineation and speed is federally controlled can be shown to have caused more than 50 percent of the erosion.

Canada: Ontario's programs

The minister of housing has issued orders for all towns to restrict shoreland development in Norfolk and Haldiman counties.

Ontario, under the Unconditional Grants Act, may contribute to the cost of emergency projects related to erosion and flooding if the costs would place an undue burden on the local tax base. For areas designated by the cabinet as Disaster Relief Areas, Ontario provides matching funds to help cover losses to year-round residences, to furnishings and equipment, farm buildings, and small business structures. Low-interest loans are available through municipalities to private owners under the Shoreline Property Assistance Act. Under this act, loans can be obtained for up to 75 percent of the cost of construction and repair of shore protection structures as well as repair of damages to buildings and structures caused by high water or ice impact.

The Ministry of Natural Resources provides on-site engineering advice on shore protection to private landowners. In addition, a shoreline section coordinator can provide sandbags for areas susceptible to flooding, and trucks can be provided to help construct or repair protective structures during emergency conditions.

Along Crown-owned shores an engineering report prepared by a professional engineer may be required before authorization to construct shore protection works is granted under the Public Lands Act.

Along privately owned shores, if the shore protection works extend below the high water mark permission is required under the Public Lands Act. An evaluation by a professional engineer is required. The evaluation must be submitted to the local district office of the Ontario Ministry of Natural Resources.

United States: federal programs

Three general U.S. federal programs are of immediate interest to shoreline property owners.

Coastal Zone Management Act

The Coastal Zone Management Act of 1972 (P.L. 92-583) was passed in recognition of the importance of the coastal zone of the United States as a national resource and the potential for adverse effects resulting from intense development pressures. The act (reference 93, appendix B) authorized a voluntary program of financial assistance to states to manage their coasts. The act states: "The key to more effective protection and use of the land and water resources of the coastal zone is to encourage states to exercise their full authority over the land and waters in the coastal zone by assisting states . . . in developing land and water-use programs . . . for dealing with coastal land and water-use decisions of more than local significance."

The state level of government has prime responsibility for achieving "effective management, beneficial use, protection, and development of the coastal zone." The requirements of the state plans are (1) that the management program be comprehensive, emphasizing important ecological, cultural, historic, and aesthetic values, and (2) that there be sufficient policies of an enforceable nature to ensure the implementation of and adherence to the management program. The Coastal Zone Management Act also stipulates that federal actions affecting the coastal zone shall be consistent with approved state management programs. Before a federal license or permit affecting land or water use within the coastal zone can be issued, the state must concur that such certification is consistent with its own coastal management program.

Individual state programs are described in the next section; however, a brief listing of participation in the program by Lake Erie states is worthy of note. The Michigan Coastal Management Program (MCMP) was one of the earliest programs by a Great Lakes state, approved in August 1978. The Pennsylvania Coastal Zone Management Program (PCZMP) has been in existence since September 1980 under the state's Department of Environmental Resources (DER). The New York State Coastal Management Program (CMP) was approved in September 1982, and its implementation is the responsibility of the Department of State along with the Department of Environmental Conservation. (See appendix A for addresses of the various state agencies.)

Ohio never joined the Coastal Zone Management Program although there were six years of program development effort. As a result, the state has not been eligible for federal funds under the CZM program—funds that sister participating states have been receiving. In spite of being a nonparticipant in the federal program, Ohio does have an interest in coastal management, developed primarily through the Ohio Department of Natural Resources (DNR). A private nonprofit corporation, the Ohio Coastal Resource Management Project (OCRMP) has continued to work for the development of policy recommendations for the future.

As coastal problems continue to develop, new or expanded management programs are likely to develop.

National Flood Insurance Program

One of the most significant legal pressures applied to encourage land-use planning and management in the coastal zone is the National Flood Insurance Program (NFIP). The National Flood Insurance Act of 1968 (P.L. 90-448) as amended by the Flood Disaster Protection Act of 1973 (P.L. 92-234) requires that the community meet certain conditions before its home owners are eligible to purchase flood insurance. The consequence of not purchasing flood insurance is that property owners who experience flood loss may not receive most forms of federal financial assistance. For example, home mortgage insurance from the Federal Housing Administration (FHA) and Veterans Administration (VA) and aid from the Small Business Administration or Department of Agriculture are available only if the individual and community involved have complied with the requirements of the NFIP. In addition, federal funds for shoreline engineering, waste disposal, or water treatment systems in the flood zones are not likely to be approved for nonparticipating communities.

Communities must adopt certain land-use measures to make flood insurance available at reasonable rates. The law applies both to river floodplains and to coastal areas that are subject to storm surges. This includes open-ocean shorelines that are heavily developed. If sea level rises or the frequency of destructive hurricanes increases, flood insurance rates will increase because they are actuarial (based on the experience of loss, which is likely to increase).

The initiative for qualifying for the program rests with the community, which must contact the Federal Emergency Management Agency (FEMA; see appendixes A and B). Once the community adopts initial land-use measures and applies for eligibility, FEMA designates the community as eligible for subsidized insurance under the emergency program. After a community qualifies for the emergency program, a Flood Hazard Boundary Map (FHBM) that delineates approximate flood-prone areas is adopted. Ultimately, FEMA conducts an engineering study and provides the community with a Flood Insurance Rate Map (FIRM), which offers more detailed delineations of flood hazard areas and data on 100-year-flood elevations. A 100-year flood is a flood having a 1 percent annual probability of occurrence. (It is not a flood that happens only once every 100 years. This misunderstanding led a Lake Saint Clair resident, told that his basement would flood in a 100-year flood, to remark that judging by the number of times his basement has flooded he is now 3,000 years old!) The FIRM shows flood insurance zones (different categories of flooding). Low-lying areas immediately next to the lake may be in V-zones, which would be penetrated by waves on top of the storm-surge flood level. Residential buildings in such a zone must meet higher elevation requirements to accommodate potential wave height, and may be charged higher actuarial rates. Nonresidential buildings should be flood-proofed in order to be eligible for flood insurance.

Before building or buying a home, an individual should ask certain basic questions: Is the community I'm locating in covered by the NFIP? If not, why not? Is my building site above the 100-year-flood level, plus wave height? What are the structural requirements for my building? What are the limits of coverage?

Most lending institutions and building inspectors will be aware of mapped flood-prone areas, but it is wise to confirm such information with the appropriate insurance representative or program office (see appendix A, under *Insurance*).

Some flood insurance facts

Is shoreline erosion covered? The answer to this question is not a clear "Yes" or "No." According to explanations in FEMA correspondence, the Flood Disaster Protection Act of 1973 extended the flood insurance program to cover losses from certain types of erosion. FEMA notes that the report of the Senate Committee on Banking, Housing, and Urban Affairs to accompany H.R. 8449 states, in part:

As the debate on the House floor made clear, such erosion coverage is appropriate for the purpose of this program only to the extent that it is flood-related; that is suddenly caused by unusually high water level in a natural body of water, accompanied by a severe storm, or by an unanticipated force of nature, such as a flash flood or an abnormal tidal surge, or by some similarly unforeseeable event. The amendment is not intended to provide coverage for losses incurred when properties built too close to shorelines are eventually damaged as the result of normal and continuous wearing away of land by ordinary wave action. If the loss is gradual and takes place over the course of years, it would not be covered.

In the Great Lakes the issue of the cause of erosion or related time of flooding is further complicated by the changing water levels of the lakes through time. This has been addressed, and FEMA notes the congressional language of statute 42 U.S.C. Section 4001(g), which states:

The Congress also finds that (1) the damage and loss which may result from the erosion and undermining of shorelines by waves or currents in lakes and other bodies of water exceeding anticipated cyclical levels is related in cause and similar in effect to that which results directly from storms, deluges, overflowing waters, and other forms of flooding, and (2) the problems involved in providing protection available through a federal or federally sponsored program are similar to those which exist in connection with efforts to provide protection against damage and loss caused by such other forms of flooding. It is therefore the further purpose of this title for purposes of the flood insurance program, protection against damage and loss resulting from the erosion and undermining of shorelines by waves or current in lakes and other bodies of water exceeding anticipated cyclical levels.

Some property owners are unaware of these requirements for a claim to be valid. In any case you must have a flood insurance policy under the program to make a claim.

Must my house or structure be in a flood zone in order to purchase flood insurance? No, any building can be insured as long as the community is a participant in the NFIP. It may not occur to an owner of property high above lake level or on an upland even to consider purchasing flood insurance; but if flash flooding, damage due to lakeshore erosion during storms during higher-than-anticipated lake levels, or runoff flooding due to exceptionally high rainfall are likely to affect the property, you may want to consider purchasing flood insurance. Flood insurance covers losses that result from temporary flooding of normally dry land, the overflow of inland waters, and the unusual and rapid accumulation of surface water runoff from any source.

Know the difference between a home owner's policy and a flood insurance policy. Flood insurance offers the flood victim a less expensive and broader form of protection than would be available through a postdisaster loan. Flood insurance is a separate policy from home owner's insurance. The latter covers only damage from wind or wind-driven rain.

Check whether your property location has been identified on the federal Flood Insurance Rate Map or Flood Hazard Boundary Map. If you are located in a flood-prone area, you must purchase flood insurance to be eligible for all forms of federal or federally related financial, building, or acquisition assistance. To locate your property on the FIRM or FHBM, see

your insurance agent. Also, keep the following points in mind:
(1) You need a separate policy for each structure. (2) If you own
the building, you can insure structure and contents, or contents
only, or structure only. (3) If you rent the building, you need
only insure the contents. A separate policy is required to insure
the property of each tenant.

*A condominium unit that is a traditional town house or row
house is considered for flood insurance purposes as a single-
family dwelling.* The individual units may be insured separately.

*Most mobile homes are eligible for coverage if they are on
foundations.* It makes no difference whether the foundation is
permanent or whether the wheels are removed either at the time
of purchase or while the home is on the foundation.

Some structures are not eligible for flood insurance. Be aware
that the following are not covered by flood insurance: travel
trailers and campers, new mobile homes in V-zones, fences,
seawalls and similar structures, retaining walls, septic tanks,
outdoor swimming pools, gas and liquid storage tanks, wharves,
piers, bulkheads, growing crops, shrubbery, land, livestock, roads,
or motor vehicles.

*One insurance broker cannot charge you more than another
for the same flood insurance policy.* Rates are set by the federal
government.

There is a five-day waiting period. Coverage will not be effective until five days after the date of application, except when
there is a transfer of title.

Expect changes in the National Flood Insurance Program.

Property owners and prospective owners should anticipate
changes in the NFIP. Already a much greater share of the cost for
the program has been shifted from the taxpayer to the property
owner as actuarial rates have taken effect, causing a dramatic rise
in premiums. This trend will continue. The rules are periodically
reviewed and revised, sometimes becoming stronger (e.g., re-
visions outlined in the *Federal Register*, September 4, 1985, part
V, Federal Emergency Management Agency). In 1982 Congress
passed the Coastal Barrier Resources Act (P.L. 97-348) to reduce
the wasteful spending of federal tax dollars for development-
related activities in coastal high-risk areas of certain barrier
islands of the open-ocean coast. Similar restrictions may be
adopted for parts of the Great Lakes to protect the water resources and minimize the loss of valuable fish and wildlife habitat
resulting from undesirable, as well as unsafe, coastal development. Restricted areas would be denied subsidies (including
flood insurance) from the federal government to build bridges,
roads, or other infrastructure such as sewer and water lines that
encourage development.

U.S. Army Corps of Engineers, jurisdiction and programs

The Federal Water Pollution Control Act Amendments of 1972
control any type of land use that generates, or may generate,
water pollution. They also regulate dredging and filling of wetlands and water bodies. Anyone who wishes to dredge, fill, or

place any structure in navigable water (almost any body of water) must apply for a permit from the U.S. Army Corps of Engineers (see appendix A). Such structures include boat docks and ramps, and shoreline stabilization structures (e.g., seawalls, bulkheads, groins, etc.). In most states the coastal zone management regulations require a similar permit, and in most cases a joint permit application is submitted through the state agency (see appendix A).

Erosion damage assistance is authorized through the Corps of Engineers if federally built navigation structures can be shown to have caused erosion. The corps is responsible for the construction and maintenance of harbor and channel protective structures (e.g., jetties, breakwaters, levees, and similar structures). They may also be the lead agency in the planning and construction of large shoreline stabilization and flood-control works such as seawalls.

Of particular interest is the Advance Measures Program, which is designed to prevent loss of life and damage to improved property from Great Lakes flooding. The program was initiated in 1985 in response to record-setting lake levels on Lakes Erie and Huron. Formal projects were developed and approved at five Michigan sites on Lake Erie (Labo Island, Milleman, Detroit Beach, Estral Beach, and Luna Pier) and at three Ohio sites (Reno Beach-Howard Farms, Whites Landing, and Bayview). Unlike the 1973 Operation Foresight Program, the Advanced Measures Program implemented shared costs. The federal government's share is limited to 70 percent of the project cost, with the remainder to be paid by state and local governments.

The U.S. Army Corps of Engineers also assists local communities during floods by providing technical assistance, supplies, and equipment.

United States: state programs

Michigan

The Department of Natural Resources (DNR) regulates some of the uses and development of coastal areas affected by erosion and flooding and of coastal areas of environmental concern through the Shorelands Protection and Management Act. The DNR has additional coastal zone management authority under the Great Lakes Submerged Lands Act, the Sand Dune Protection and Management Act, the Inland Lakes and Streams Act, and the Michigan Environmental Act. Building setbacks are required in areas eroding at one foot per year or greater (high-risk erosion areas) and are generally planned to protect a permanent structure for 30 years.

A flood risk area is an area that is in the 100-year floodplain (an area that in a given year has a 1 percent chance of being flooded). Building requirements must be met and permits obtained for building in these areas.

Shore structures are regulated by the Department of Natural Resources. Permits are required for the construction of structures such as seawalls, groins, and breakwaters, and permit applications are processed jointly through the U.S. Army Corps of Engineers. In addition, the law controls such activities as dredging and filling of submerged lands below the ordinary high water mark of 571.6 feet (International Great Lakes Datum).

The Michigan Department of Natural Resources (MDNR), with help from the Michigan State Housing Development Authority, assists home owners under the Emergency Home Moving Loan Program. Under this program lakeshore owners whose homes are in "imminent danger of destruction" from shore erosion must request an inspection of their property from the MDNR. If the inspector decides that serious damage or destruction from storm and wave action can be anticipated in the next 12 months and if the homeowner is able to obtain a loan to move the home, then the interest rate on the loan will be lowered 3 percentage points below the rate that the building owner has agreed to pay the private lending institution. Conditions for moving the structure are:

1. The building must be moved landward at least 30 feet; or if the property is within a designated high-risk erosion area under the Shorelands Protection and Management Act, it must be moved the minimum required setback distance.
2. All construction materials, including the old foundation and septic system must be removed and properly disposed of or reused as part of the moving project.
3. The building must be relocated within 90 days of the date of the loan.

If the loan is for more than $25,000 there is no subsidy for the amount of the loan over $25,000.

To apply, write:

Emergency Home Moving Program
Michigan Department of Natural Resources
Division of Land Resource Programs
P.O. Box 30028
Lansing, MI 48909

State in your letter the location of the building (include a map if possible), how far it is from the active slumping or erosion, and your intention to move it. Please include your mailing address and a daytime phone number. The building will be certified by the Division of Land Resource Programs as meeting the requirements of the program. You may contact the Lansing office at (517) 373-1950.

Ohio

In Ohio, coastal zone management of shore erosion is primarily up to the local governments, with state agencies available for assistance. For example, the city of Conneaut has adopted a building setback requirement for permanent structures built along the Lake Erie shore. The state does not have a building setback requirement. There is no state regulation of flood-risk areas; however, the Ohio Department of Natural Resources coordinates the National Flood Insurance Program.

Shore structures are regulated by the Department of Natural Resources. Permits are required by the state for the construction of structures such as seawalls, groins, and breakwaters. There is a provision for financial assistance by the state for shore erosion projects on publicly or privately owned property. Municipal financial assistance can be provided for shore erosion projects undertaken by several residents along a stretch of shore.

Submerged land is regulated by the Department of Administrative Services. The department controls activities such as the building of structures into the lake.

Pennsylvania

Pennsylvania, like Michigan, has a setback requirement. The Department of Environmental Resources calculates the average annual erosion rate for the given municipality; this rate is then multiplied by 50 for residential homes, 75 for commercial structures, and 100 for light and heavy industrial structures to determine the setback distance. Regardless of the rate, the minimum bluff setback distance is 50 feet. In addition, eight of the nine municipalities along the Pennsylvania shore (the city of Erie is shielded from Lake Erie by Presque Isle, and undergoes little shore erosion) have adopted bluff setback ordinances. The ordinances are enforced by the municipalities and are monitored by the Department of Environmental Resources (DER) on a yearly basis. An encroachment permit is required from the DER (Bureau of Dams and Waterway Management) for the construction of

shore structures. A permit is also required from the U.S. Army Corps of Engineers.

New York

The Department of Environmental Conservation is charged with the delineation of erosion hazard areas, and the municipality in which the erosion hazard area occurs is charged with the regulation of that hazard area. If the municipality does not meet the obligation, then the county is charged with regulation, and if the county does not meet the regulation, the department will regulate the erosion hazard area(s). In essence, regulations should "minimize damage to property, and . . . prevent the exacerbation of erosion hazards." For guidance in acquiring the necessary building permits and interpreting regulations, contact the Department of Regulatory Affairs in the New York State Department of Environmental Conservation (see appendix A). A permit from the Department of Environmental Conservation is required for the construction of structures such as seawalls, groins, and breakwaters. A permit is also required from the U.S. Army Corps of Engineers.

6 Reviewing the alternatives

The history of Lake Erie and its shores is one of change: natural change and human cultural change. Our views, attitudes, and philosophies with respect to occupying and using the shore are no more constant than is lake level. Beaches have been lost and property stands threatened by today's coastal hazards because development was not designed to be flexible with nature. Wide-spread fixed shoreline structures retard erosion, but at a signifi-cant cost. One need only remember that the early settlers used stretches of Lake Erie beaches as roadways. How far could you drive a wagon along Lake Erie's beaches today?

The lake and its shore are a resource to be used, enjoyed, and protected. Individually, and collectively, the purpose of our presence at the shore is to take advantage of this unique resource. We come to recreate in a variety of styles, to enjoy the aesthetics and ambience of the setting, and to use the waterways. No one settles on the lakeshore to be threatened by coastal hazards, or to be frustrated by the need to combat nature. No one builds a home or business with the intention of losing his investment or spending exorbitant sums to protect it. Yet these are often the results of shoreline settlement today.

If historic trends hold, shoreline development will continue, and many property owners will opt for stabilization structures when the inevitable erosion problems arise. But, as noted early in the book, such structures contribute to a loss of sand and narrower beaches. A day could come when most of the Lake Erie shore will lack beaches and waves will lap directly onto a shore of concrete and steel. Such a future is not a pleasant prospect.

Unfortunately, the current "solutions" to the erosion problem are in the nature of Band-Aids. No cure is being provided. The purpose of this book is to suggest that genuine consideration be given to avoiding similar problems in the future. The theme is prudence in planning and action—easy to say but not easy to implement.

Any use of Lake Erie comes at a price. A few years ago the most apparent price was water pollution, a problem particularly severe around the larger cities, but affecting water quality throughout the lake. Although pollution has not gone away, progress has been made in reducing pollution through actions ranging from the international level to the individual level. As this book goes to press (1986), Lake Erie is at its highest level in this century and thousands of home owners believe that their property and quality of life is threatened. The time is right for the same spirit of renewal that has gone on before. The range of actions again must be from international to individual.

International, provincial-state, and community actions

The most realistic approach to alleviating the natural hazards of erosion and flooding seems to be through strict regulation (see chapter 5). The future will see an increase in restrictions on shoreland use. Within limits, Lake Erie is a system that can be partially managed. Regulation and management programs both require collective actions, usually at higher governmental levels.

Reducing coastal hazards through management programs must take into account how the lake operates as a whole, and how component systems within the lake function (chapter 2). For example, actions to reduce shore erosion must consider lake level changes, regional longshore drift patterns, regional shoreline types, and sand supplies (chapter 4). Four types of collective action are suggested.

Lake level regulation

This is a complex long-term international approach to reducing flood and erosion problems. The goal is to maintain Lake Erie's water level within specific limits as is being done in Lake Ontario. Such a system presumably would reduce the extreme low levels that create serious water supply and navigational problems, as well as extreme high levels that create serious erosion and flooding problems (fig. 6.1). The reduction of high levels would reduce erosion, but not eliminate it. Big storms, reduced sand supplies, and human intervention in the dynamic shore system will result in continued erosion even during lower lake levels. Nature is the ultimate control, but engineered water diversions can modify lake level.

Regulation of the Great Lakes water levels was considered carefully in 1973 and 1981 studies published by the International Joint Commission (references 100 and 101, appendix B). Interestingly enough, the impetus behind the 1973 study was the low lake levels of the mid-1960s. The 1981 study focused on Lake Erie, largely as a result of the problems caused by shore erosion and flooding during the high lake levels of the early 1970s. The 1981 study concluded that Lake Erie's level could be modified by control structures near the head of the Niagara River. The study also concluded that such controls could not be justified because the projected economic losses to commercial navigation, recreational boating, and hydroelectric power offset the projected economic gains to recreational beach interests and lakeshore property. Unfortunately, this conclusion has led to the idea that Lake Erie is being held artificially high, and various rumors to this effect circulate sporadically in the along-the-shore communications drift.

By 1986 lakefront home owners' associations from Lakes Michigan, Huron, and Erie argued that property losses due to erosion and flooding, costs for private stabilization structures, and tax losses due to decreasing property values warranted immediate steps to reduce lake levels. Extreme views ranged from holding more water back in Lake Superior to significantly increasing the flow down the Niagara River. For example, when Lake Superior (a "regulated" lake) was at its maximum legal level and home owners on that lake were calling for lowering its level, property owners on Lakes Michigan and Huron wanted more water held back. To paraphrase one Lake Michigan prop-

6.1   Storm damage, Luna Pier, Michigan. Photograph courtesy of
The Detroit News.

erty owner's desperate view: the erosion problem could be shared. Residents of New York and Ontario should be aware that some folks in Michigan might be more than willing to sacrifice some of the beauty of Niagara Falls if it means saving their property.

Because the lakes and their uses are continually changing, reevaluation of lake level modification through various engineering works should be required. Naturally, great caution is in order before embarking on such mega-projects. Suppose for example, the decision to build structures or establish diversions *into* Lake Erie had been set into law and effect to compensate for the low water of the mid-1960s. The result could have been erosion of even greater severity in the 1970s and 1980s!

The American southwest might provide a market for Great Lakes water, and pipelines have been proposed. Any removal of water is attractive during high lake level intervals. Imagine, however, a water supply, mandated by law, going out of the lakes during successive years of drought and low lake levels. Would the citizens of Lake Erie's coastal municipalities appreciate lowered water quality for the sake of west Texas's cotton farmers or Arizona's industry?

The ultimate complexity and impediment to lake level regulation lies within the basins of the middle lakes. The interests of two countries, seven states, and one province may be so diverse that all actions will be limited compromises. Private and municipal interests are just as diverse. The lobbying of property owners is likely to be offset by similar efforts from contrary special interest groups. In any case, nature is almost certain to move

faster than political or regulatory bodies.

The bottom line is that significant engineering regulation of the Great Lakes to control the level of Lake Erie is not a likely management alternative for the immediate future.

Do-nothing alternative

This approach, letting nature take its course, may seem unworthy of mention. In truth such an approach, in modified form, is becoming increasingly attractive in many coastal states. This alternative has been mandated into law in at least two states with ocean shorelines. North Carolina and Maine now prohibit construction of fixed or "hard" structures along their coasts. Beach replenishment is permitted, but the basic thrust of the do-nothing laws is that sooner or later the buildings next to the shore will have to be moved back or allowed to fall into the sea.

Justification for this state-level approach is the well-founded realization that shore stabilization structures degrade or destroy the beach, cause offshore steepening, and ultimately allow greater wave energy to attack the shore. As noted in chapter 3, because of these associated changes, bigger and better structures are continually required, at an ever-increasing cost to the shoreline owners or community. Portions of the New Jersey shore have reached the stage where the cost of shore protection structures has exceeded the value of the property being protected.

Recently a panel of planners, engineers, geologists, and environmentalists produced a white paper on the future of American recreational beaches. Although the document, known as the "Skidaway Position Paper," was intended to apply to ocean shorelines, the general ideas are applicable to Lake Erie. The paper's summary reads: "The American shoreline is retreating. We face economic and environmental realities that leave us two choices: (1) plan a strategic retreat now, or (2) undertake a vastly expensive program of armoring the coastline and, as required, retreating through a series of unpredictable disasters."

The do-nothing alternative is a step in the direction of an orderly retreat from the shore, or a move toward a natural solution. Michigan has moved in this direction with its program to provide assistance for owners of threatened houses to move their buildings back from the shore. The alternative is the current disorderly, expensive retreat with an associated loss of the natural environment and setting.

Beach Nourishment

This intermediate-level approach is in keeping with natural processes, and is often adopted by municipalities, or by groups of home owners. As noted in chapter 3, beach nourishment is not without its problems, but if well planned and well executed, nourishment imitates the natural system.

The primary requirement for this approach is the understanding that continuous maintenance will be required. Nourishment is not a one-time action. Costs of major replenishment projects are on the order of $1 million per mile. Moreover, the cost of nourishment will increase as the supply of sand diminishes.

Sand could be dredged from navigation channels and areas of accumulation adjacent to major harbor jetties and placed in downdrift beaches. Some of the sand dredged for channel maintenance has been deposited offshore or in diked disposal areas. Such sand has been lost for human use, by being taken out of the natural drift system. Unless polluted, this resource could be used for beach maintenance.

Beach nourishment projects benefit areas beyond the artificial beach. Sand moves downdrift and offshore to feed other beaches and protective offshore bars. Sand from both upland and lake-bottom sources has been used in projects at Lakeview Park, Lorain, Ohio, and Presque Isle, Erie, Pennsylvania.

Stabilization through the construction of shore protection structures

This has been a common approach adopted by communities and on a smaller scale by property owner associations. Typical structures include riprap seawalls, breakwaters, and groin fields, sometimes used in combination with the emplacement of sand (beach nourishment). Large municipalities must arm and maintain their shores, especially harbor and port facilities, where the benefits justify the high cost.

In most cases, large structures are very expensive and are beyond the means of small communities. Federal and state economic support is necessary for such construction, usually by the

U.S. Army Corps of Engineers. Once in place, the structure must be permanently maintained, literally forever. The long-term economic obligation should give any community pause when reviewing stabilization structures as an option. Add to this the known impacts of seawalls, breakwaters, and groins on the beach system (see chapter 3), and the conclusion should be that it is a "last resort" option, and a systematic retreat may be in order.

On a somewhat smaller scale, groups of adjacent property owners sometimes build intermediate-sized structures of the same type noted above. These structures are larger than the walls or few groins an individual homeowner might install, but not as massive, or elaborate as an Army Corps project. Nevertheless, such structures are expensive to build, and they must also be maintained. If you belong to an association that is considering a stabilization structure you should examine your own willingness to take on the financial obligation for continued contributions to maintenance. Furthermore, will the initial investors in the structure all stick together in their commitment to finance the structure's future maintenance? The same questions apply when a community considers a stabilization structure. Is it economically feasible in the future as well as now? Does the community understand the harmful side effects of the structure (e.g., interruption of longshore drift, offshore steepening, loss of beach)?

Where structures are now in place the die is cast, but even communities that have yet to experience problems should not be complacent. A successful strategist always plans the options for retreat. Moving back or even planning future nourishment is preferable to waiting for a crisis to develop and then seeking the "quick fix".

For areas where the erosion problem is arising, shoreline stabilization structures might be placed with a plan. For example, artificial headlands, although not in keeping with a true natural system, may offer temporary respite from erosion along heavily developed shorelines. Such a design might be considered by a municipality or property owners' association. In this alternative, relatively short (a few hundred feet) sections of shore are stabilized by shore protection structures and landscaping to retard erosion (these structures prevent the formation of a beach, however, because of wave reflection from the structures and accelerated longshore transport). These short, protected sections of shoreline are separated by unprotected sections, which are allowed to erode to form embayments of several tens to a few hundred feet in depth, depending upon the length of the adjacent protected stretches (the longer the protected stretches, the deeper the embayments). The embayed areas trap sand to form pocket beaches because the waves and longshore currents do not have the strength to move the sand out of the embayments (fig. 3.10). If sufficient sand is not available to make a beach along the embayed shore, sand is trucked in to construct an artificial beach. Houses near the beach are still in the hazard zone, but erosion is reduced because of the more sheltered location on the pocket beach. Lakeshore owners still have a beach for recreation; however, some property owners have to be willing to sacrifice some of their land for the embayment to form.

Along the present Lake Erie shore, some of these headland and embayment areas already have formed as a consequence of unplanned shore protection. Some property owners have stabilized their shore, but others are not willing or not able to build structures. If such plans were mandated, they would involve a trade-off between shore protection and land loss. Eventually, a beach would protect part of the shore from waves. Once this occurred, the embayed shore slope would approach a stable configuration, especially if it were landscaped. Naturally the shore protection structures that front the headlands would need to be maintained to keep their effectiveness. The latter aspect is the weakness of such a plan because the stabilized headland would experience all of the problems associated with stabilization structures. Owners of designated headland property, and those who would have to keep paying for headland maintenance, should be skeptical of such a plan because it is a last-resort method of questionable cost and duration.

Individual actions

A property owner might hope that governmental agencies will end the erosion problem by miraculously lowering lake level, building and nourishing protective beaches, or taking some other major action. In truth, such actions are often impossible, may be impractical because of high costs, or are proscribed because of conflicting land-lake uses or political-social-economic impedi-

ments. When such steps are practical, long delays are certain between the critical time when erosion is occurring and the time of implementation. Major actions require legislation for approval and funding, plus the time required for regional studies to evaluate expected results, cost-benefit ratios, and environmental impact. More time is required to carry out the beach nourishment or construct the engineering works.

As a result, some property owners feel that they are running out of time. The pressure is to take immediate action to protect a house, building, or other property.

To avoid possible rash, expensive actions, the prudent property owner should plan ahead, avoid coastal hazards where possible, and evaluate all alternatives. The process begins by educating yourself in regard to coastal processes, and weighing the benefits of living at the shore against risks and potential losses. First, review your use and expectations of coastal property. For example, do you prize a recreational beach, an uncluttered shoreline, and a setting that approximates natural conditions? Are you wealthy enough to hold back nature individually if you choose to combat erosion? Are you part of a larger community that will choose to take collective action, whether or not you agree with such action? Such actions might include petitioning state-provincial or federal governments to build stabilization structures or initiate beach nourishment projects. What will happen to your property in such a project, or if your neighbor builds a structure? What will happen to your neighbor if you build such a structure? If you intend to sell the property in the future, what will potential buyers be looking for in shoreline property? What are your expectations in regard to property values, and are such expectations realistic given the history of shoreline change?

After reviewing both your needs and the reality of nature, take prudent steps for the future. Consider the following as general guidelines in a descending order of priority:

1. *Pick a safe site initially.* When locating at the shore follow the suggestions given in chapter 4. Choose a site where the erosion rate is known to be low, or where accretion is occurring. Look for a shore of resistant or stable material (e.g., bedrock rather than unconsolidated sediment). Avoid shorelines where evidence of erosion exists, e.g., the presence of shore protection structures, fallen trees and downed vegetation, bare ground or scars of slumps and slides, and wave-cut scarps. Look for signs of stability such as a healthy ground cover of vegetation, and a wide beach. Make sure the elevation of the site is above the 100-year flood level. If not, then any existing or proposed residence should have its first floor above the 100-year flood elevation.

2. *Make sure the house or building is set well back from the shoreline.* In Michigan the required setback for new structures is a distance from the shore equal to thirty times the average annual erosion rate. The requirement is intended to guarantee the existence of most houses through their initial mortgage, and this setback should be regarded as a minimum. Average erosion rates are based on relatively short-term records, and may increase if lake level remains high. Even if the average is accurate, most owners expect their property to last longer than thirty years.

Older houses of nearly a century in age remain along the shore today because they were originally set well back from the shore. Ultimately, though many of these older houses are claimed by the lake.

3. *Elevate and floodproof your house or building.* A house may be relatively safe from shoreline erosion, but still located within a flood zone (fig. 6.2). If you choose to build within such a hazard zone, make sure your house is elevated so that the first dwelling floor is above the 100-year flood elevation. References 81 through 83 in appendix B provide construction guidelines for buildings within a flood zone. Existing buildings can be strengthened and floodproofed (appendix B, references 84 through 91).

4. *Consider moving your house or building.* The cost of moving a structure may be less than the ultimate cost of stabilizing a site. Decide if you simply want to move your house farther away from the lake, or move it to another area. Long-term erosion rate data can help you decide how far back from the lake you should move. Remember that erosion rate information is accurate for the past, but is only a guide for the future. Therefore, consult with engineers, scientists, or governmental advisory agencies (appendix A) in order to determine the best plan of action. If, for example, your structure is located in an embayment and fronted by a beach, the erosion rate may have decreased significantly enough below the average so that a move is unnecessary. On the other hand, if there is really no alternative to moving, then hire a contractor whose business is moving homes, and let him do it. From both an aesthetic and an environmental

standpoint, the best approach is to move threatened buildings away from the shore and let nature take its course. When buying an existing older house that may be threatened in the future, consider the type of structure and whether it will be easy to move. For example, a wood-frame house is easier to move than a cinder-block house. Check programs (see chapter 5) that may provide some financial assistance to move a residence.

5. *Do nothing.* Although this suggestion may seem inappropriate and unreasonable, letting the house fall in may be a more appropriate and less expensive action than fighting nature. Evaluate the costs of moving back, the age-condition-financing of the house in question, and the real costs of shoreline stabilization structures (i.e., initial cost, maintenance, replacement). Is it possible that insurance (see next item) will cover part of the loss if the do-nothing approach is followed? Consider also that you will be responsible for removing the structure from the lake after it has fallen in the water.

6. *Carry flood insurance.* If you live in a flood zone, if your house is threatened by shoreline erosion, or if it is likely to be flooded as a result of storms during times when normal cyclic high lake levels are exceeded, insurance is probably a worthwhile

6.2 Lake Erie storm surge aftermath, Luna Pier, Michigan. Photograph courtesy of *The Detroit News.*

expense. Note that you do not have to live in a flood zone to be eligible to obtain flood insurance. Your community must, however, be a participant in the National Flood Insurance Program (see chapter 5).

7. *Rebuild or nourish your beach.* Beach nourishment in front of individual properties is sometimes attempted. Such action provides only temporary protection, and then only under conditions where some natural protection already exists and only under low-energy wave conditions, i.e., where the sand is not likely to migrate away. Such conditions might apply in an embayment, which is protected by adjacent headlands and is the end point for beach drift. Beach nourishment, however, is usually inappropriate, expensive, and of little lasting protection for individual property owners. Building beaches is a large-scale (hundreds of yards to miles) approach to combating erosion.

8. *As a last resort, build stabilization structures.* As noted in chapter 3, stabilization structures (i.e., breakwaters, seawalls, groins) usually cause as many problems as they solve. We recommend that this approach be considered only as a last-resort solution to shoreline erosion. Remember that in most cases you will cause degradation of your beach. Remember also that the cost of long-term protection of the shore is beyond the financial means of many shoreline dwellers.

For those who choose the last-resort method

Just as coastal development will continue, communities and individual property owners will continue to opt for stabilization structures. Protecting the shore with such structures is the most commonly tried solution where the shore is eroding fairly rapidly (1 to 2 feet per year or more), and where buildings are close to the shoreline. Aside from meeting legal requirements (see appendix A) there are several things that a landowner can do in a successful shore protection project. From the planning perspective, initially you should talk to your neighbors to find out what they have done. See if their efforts have been successful and if they might be interested in a group project. Contact engineers and scientists from the various federal, provincial, and state governments for free advice (see appendix A for a list of agencies willing to help with erosion problems). These people should be able to give you a pretty good idea of what you need to do to protect your property, and how much it will cost. Once you have an idea of what has happened in the past and what is presently happening, you can predict what is likely to happen in the future. You can then decide whether to go ahead and armor the shore. Even if you decide that you don't want to put in shore protection structures, you still need to have a good idea of how fast erosion will progress, and how long it will be before your house collapses into the lake. Alternatively, if you do decide to go ahead and try to stabilize the shore, follow these guidelines:

1. *Decide exactly what you want to do.* For example, do you simply want to protect the toe of the slope from waves or do you want to conserve the beach as well? If you want a beach, be aware of the immediate source of the sand for the beach. If the

main source is longshore drift and you or a neighbor installs groins that alter the flow of sand, then the downdrift beach may narrow or disappear. Remember, most shore protection structures will affect the adjacent beach or shore, yours or your neighbors', beyond the immediate position of the structure.

2. *Try to coordinate your project with your neighbors.* If the shore is already built up, you probably should tie in your structure with those of your neighbors. Alternatively, if several adjacent owners want to have work done, try to have a single plan to reduce costs and construct a more coherent structure. Furthermore, if the stretch is long enough and conditions warrant, your corporation or conservancy district might want to try a beach nourishment project. You might be able to apply for public funds as a quasi-governmental agency ("erosion control district").

3. *If you aren't going to do the work yourself, find an experienced lake contractor to do the work.* Look at some of the contractor's completed projects with him or her and ask previous clients about the success of the projects.

4. *Obtain a permit* (see appendix A).

5. *Once the structure is built, be prepared for continuing maintenance expense.* Structures should not be considered permanent. Waves, ice, and slope forces eventually modify or destroy most structures, often in a fairly short period of time. Costs to build the structures vary greatly. Along low relief, sheltered stretches of shoreline, the costs can be a few tens of dollars per foot. Along high relief, exposed stretches of shoreline the costs are a few hundreds of dollars per foot. Naturally, the better-built, the more massive, the more long-lasting the structure, the greater the cost.

6. *If you do the work yourself, utilize proven general guidelines such as those found in the U.S. Army Corps of Engineers booklet, "Help Yourself"* (reference 71, appendix B). This booklet lists the following six construction and maintenance rules: (a) provide adequate protection for the toe of the structure so that it will not be undermined; (b) secure both ends of the shore protection works against flanking; (c) check foundation conditions; (d) use material that is heavy enough that waves will not move individual pieces of the protection; (e) build the revetment high enough that waves cannot overtop it (spray overtopping is all right, but not "green water"); and (f) make sure that voids between individual pieces of protection material are small enough that underlying material is not washed out by waves. The use of filtercloth is recommended.

7. *Utilize landscaping, plantings, and drain systems to stabilize slopes in back of shore protection structures.* Although the ultimate problem is the waves, shore protection built at the toe of a slope does not prevent slope failure. It is particularly disheartening to see an apparently stable shore undergo major slope failure after the installation of costly stabilization structures. Slopes often fail even in areas where there are no signs of past slope failure. Therefore, if you have the room, grade or terrace and drain the slope to help prevent failure that may damage or destroy your shore protection structure as well as property at the top of the slope. Plantings of grass and other vegetation may also

stabilize the slope (references 61, 66, and 75, appendix B).

8. *Don't be taken in by so-called new, sensational, or innovative devices that are sold with the promise of stopping erosion.* In recent years a variety of structures have been promoted under catchy names. These structures are essentially modifications of standard seawall, groin, breakwater, or other structural designs. In some cases the only apparent difference is the construction material of the structures, or the position of emplacement. Typically hailed as successful after one season (usually a quiet summer), the design receives glowing press reports. But these wonders are soon forgotten as they fail under the onslaught of high-water storms. In the meantime, it is the property owner who has been sandbagged instead of the beach. When choosing a structure, consider that engineers have been designing structures since the early days of development. Seek the advice of experts as noted above, and look for examples of structures that have stood up to waves and flooding for a reasonable period of time.

All actions, whether international, federal, provincial, state, local, or individual are influenced by public participation. The best way to contribute to the solutions of shoreline problems is to

6.3A   Lake Erie storm surge, Gibraltar, Michigan. Photograph courtesy of *The Detroit News*.
6.3B   Lake Erie erosion, Willoughby, Ohio. Photograph by Ed Suba, Jr., courtesy of *Akron Beacon Journal*.

participate in the coastal management process. Learn about this great resource, Lake Erie, and know the regulators as well as the regulations. Join in community planning groups, attend public hearings, become involved in the wide range of groups that focus their attention on conserving the lake and its associated environments. All points of view must be heard and a balanced multiple lake use reached. Public apathy could be the greatest hazard of all if it results in poor coastal management.

Obey nature

Erosion may be prevented in the short run by shoreline structures or beach nourishment; however, these methods are costly and long-term—definitely not one-time solutions. The generations of structures that front the present shore provide mute testimony that erosion is difficult to stop in spite of some of the best human efforts. Moreover, the effects of such structures on immediate and adjacent shores tell us that the long-term price for slowing shore erosion is the loss of the beach, the loss of aesthetics, and the loss of certain qualities that were the reason for living on the shore in the first place. When choosing to go the construction route to fight shore erosion, be prepared for the expense of not only an engineering structure, but of protective landscaping, and the continued maintenance of both. Also be prepared for the lake's ultimate victory (fig. 6.3).

Appendix A

A guide to federal, provincial, and state
agencies involved in coastal development

Numerous agencies at all levels of government are engaged in planning, regulating or studying coastal development. These agencies issue permits for various phases of construction and provide information on development to the home owner, developer, or planner. Following is an alphabetical list of topics related to coastal development; under each topic are the names of agencies to consult for information on that topic. Some of these sources also provide information for noncoastal areas. For a more comprehensive listing of agencies obtain a copy of *The Great Lakes Directory* from the Center for the Great Lakes, 435 North Michigan Avenue, Suite 1733, Chicago, IL 60611; phone (312) 645-0901.

In general, the best agencies to contact for most problems and questions are the state departments of natural or environmental resources, the state Sea Grant institutions, and the district offices of the U.S. Army Corps of Engineers. If they don't have the information you need, they can usually direct you to someone who can help. In some cases, the best information sources are the field offices of federal and state agencies. These agencies include:

Michigan

U.S. Army Corps of Engineers
Detroit District
P.O. Box 1027
Detroit, MI 48231
Phone: (313) 226-6413

The Office of the Great Lakes
7th Floor, Executive Division
Michigan Department of Natural Resources
Stevens T. Mason Building
Lansing, MI 48909
Phone: (517) 373-3588

Michigan Department of Natural Resources
Great Lakes Shoreland Section
P.O. Box 30028
Lansing, MI 48909
Phone: (517) 373-1950

Michigan Sea Grant College Program
The University of Michigan
2200 Bonisteel Boulevard
Ann Arbor, MI 48109
Phone: (313) 763-1437

District Sea Grant Agent
Michigan Cooperative Extension Service
County Building, 11th Floor
Mount Clemons, MI 48043
Phone: (313) 469-5180

Ohio

U.S. Army Corps of Engineers
Buffalo District
1776 Niagara Street
Buffalo, NY 14207
Phone: (716) 876-5454

U.S. Army Corps of Engineers
Toledo Projects Office
P.O. Box 5002
Toledo, OH 43611
Phone: (419) 259-6480

U.S. Army Corps of Engineers
Cleveland Projects Office
Foot of East 9th Street
Cleveland, OH 44114
Phone: (216) 522-4957

Ohio Department of Natural Resources
Division of Geological Survey
P.O. Box 650
Sandusky, OH 44870
Phone: (419) 626-4296

Ohio Sea Grant Program
Ohio State University
484 West 12th Avenue
Columbus, OH 43210
Phone: (614) 422-0557

District Sea Grant Agent
Ohio Cooperative Extension Service
99 East Erie Street
Painesville, OH 44077
Phone: (216) 357-2582

Pennsylvania

U.S. Army Corps of Engineers
Buffalo District
1776 Niagara Street
Buffalo, NY 14207
Phone: (716) 876-5454

U.S. Army Corps of Engineers
Cleveland Projects Office
Foot of East 9th Street
Cleveland, OH 44114
Phone: (216) 522-4957

Pennsylvania Department of Environmental Resources
Division of Coastal Zone Management
P.O. Box 1467
Harrisburg, PA 17120
Phone: (717) 783-9500

Erie County Department of Planning
Erie County Courthouse
Erie, PA 16501
Phone: (814) 452-3333

New York

U.S. Army Corps of Engineers
Buffalo District
1776 Niagara Street
Buffalo, NY 14207
Phone: (716) 876-5454

U.S. Army Corps of Engineers
Cleveland Projects Office
Foot of East 9th Street
Cleveland, OH 44114
Phone: (216) 522-4957

New York Sea Grant Institute
37 Elk Street
Albany, NY 12246
Phone: (518) 436-0701

District Sea Grant Agent
New York Sea Grant Extension Program
21 South Grove Street
East Aurora, NY 14052
Phone: (716) 652-5453

New York Department of State
Coastal Management Program
162 Washington Avenue
Albany, NY 12231
Phone: (518) 474-3643

Ontario

Environment Canada
Inland Waters Directorate, Ontario Region
Water Planning and Management Branch
P.O. Box 5050, 867 Lakeshore Road
Burlington, Ontario L7R 4A6
Phone: (416) 336-4712

Ontario Ministry of Natural Resources
Conservation Authorities and Water Management Branch
Whitney Block, Room 5620
99 Wellesley Street West
Toronto, Ontario M7A 1W3
Phone: (416) 965-6287

Appendix B presents a list of references, some of which provide additional agency listings as well as basic information of interest to the coastal dweller.

Aerial photography and remote-sensing imagery

Persons interested in aerial photography, remote-sensing imagery, or agencies that supply aerial photographs or images should contact the appropriate office listed below.

For historic listing of available photography (type, scale, year flown, coverage, percentage of cloud cover, etc.) contact:

National Cartographic Information Center
U.S. Geological Survey
507 National Center
Reston, VA 22092
Phone: (703) 860-6045

Energy, Mines and Resources, Canada
National Air Photo Library
615 Booth Street
Ottawa, Ontario K1A 0E9
Phone: (613) 995-4560

Recent aerial photography is available from:

Energy, Mines and Resources, Canada
National Air Photo Library
615 Booth Street
Ottawa, Ontario K1A 0E9
Phone: (613) 995-4560

Ontario Ministry of Natural Resources
Public Information Centre
Aerial Photographs
Whitney Block Room 1640
99 Wellesley Street West
Toronto, Ontario M7A 1W3
Phone: (416) 965-1123

U.S. Department of Agriculture
Agricultural Stabilization and Conservation Service
Aerial Photography Field Office
2222 West, 2300 South
P.O. Box 30010
Salt Lake City, UT 84125

Request "status of aerial photography coverage" for the individual state. Black-and-white vertical aerial photos are available for coastal counties. Black-and-white vertical aerial photographs of the lakeshore are also available from:

U.S. Army Corps of Engineers
Detroit District
P.O. Box 1027
Detroit, MI 48231
Phone: (313) 226-6413

U.S. Army Corps of Engineers
Buffalo District
1776 Niagara Street
Buffalo, NY 14207
Phone: (716) 876-5454

Michigan Department of Natural Resources
Land Resources Programs Division
P.O. Box 30028
Lansing, MI 48909
Phone: (517) 373-3328

Ohio Department of Transportation
Aerial Engineering
1600 West Broad Street
Columbus, OH 43216
Phone: (614) 275-1359

Ohio Department of Natural Resources
Division of Soil and Water Conservation
1939 Fountain Square Court
Columbus, OH 43224
Phone: (614) 265-6769

New York Department of Transportation
Mapping Services Bureau
State Office Campus, Bldg. 5
Albany, NY 12232
Phone: (518) 452-4408

Other, more conveniently located sources may have aerial photographs for your area of interest available for inspection. These include the office of the tax assessor in your county, departments of geology or geography in local colleges and universities, regional planning agencies, the district offices of the U.S. Army Corps of Engineers, and the Departments of Natural Resources for the different states.

For information on high altitude photography or satellite imagery contact:

EROS Data Center
U.S. Geological Survey
Sioux Falls, SD 57198
Phone: (605) 594-6511

Energy, Mines and Resources, Canada
Canada Centre for Remote Sensing
Prince Albert Satellite Station
P.O. Box 1150
Prince Albert, Saskatchewan S6V 5S7
Phone: (306) 764-3602

Ontario Ministry of Natural Resources
Ontario Centre for Remote Sensing
880 Bay Street
Toronto, Ontario M5S 1Z8
Phone: (416) 965-8411

Archives and records

Historic information on coastal counties and possible sources for historic maps and photographs:

Michigan Department of State
History Division
208 North Capitol Avenue
Lansing, MI 48918
Phone: (517) 373-0510

The Ohio Historical Society
1982 Velma Avenue
Columbus, OH 43211
Phone: (614) 466-1500

Pennsylvania Bureau of Archives and History
William Penn Memorial Museum and Archives Bldg.
P.O. Box 1026
Harrisburg, PA 17120
Phone: (717) 787-3051

State Archives
New York State Education Department
10A46 Cultural Education Center
Empire State Plaza
Albany, NY 12230
Phone: (518) 474-1195

Archives of Ontario
77 Grenville Street
Queen's Park
Toronto, Ontario M7A 2R9
Phone: (416) 965-4039

Bridges and causeways

The Coast Guards have jurisdiction over the issuing of permits to build bridges or causeways that will affect navigable waters. Information is available from:

Commander, 9th Coast Guard District
1240 East 9th Street
Cleveland, OH 44199
Phone: (216) 522-3910

Transport Canada
Canadian Coast Guard
Navigable Waters Protection Act Program
Prescott Coast Guard District
P.O. Box 1000
Prescott, Ontario K0E 1T0
Phone: (613) 925-2865

Building codes and zoning

Most communities have adopted comprehensive plans in terms of zoning and building codes. The BOCA building code is widely used in the Great Lakes region in the United States. Check with your county or city building department for permitted uses and building code. If your structure is in the 100-year flood zone and you intend to obtain flood insurance, certain minimum elevation and floodproofing requirements must be met. (See *Insurance*).

Erie County Dept. of Planning
Erie County Courthouse
Erie, PA 16501
Phone: (814) 452-3333

New York Department of Environmental Conservation
Department of Regulatory Affairs
University of New York, Bldg. 40
Stony Brook, NY 11794
Phone: (516) 751-7900

Canada Mortgage and Housing Corporation
London Branch Office
P.O. Box 2845
London, Ontario N6A 4H4
Phone: (519) 438-1731

Windsor Branch Office
P.O. Box 240 Station "A"
Windsor, Ontario N9K 6K7
Phone: (519) 256-8221

Hamilton-Burlington Branch Office
350 King Street East Room 202
P.O. Box 56
Hamilton, Ontario L8N 3B1
Phone: (416) 572-2451

Ontario Mortgage Corporation
A Subsidiary of Ontario Land Corporation
777 Bay Street 15th Floor
Toronto, Ontario M5G 2E5
Phone: (416) 585-6015

County of Essex Planning Board
P.O. Box 1570
360 Fairview Avenue West
Essex, Ontario N8M 1Y6
Phone: (519) 351-1010

County of Kent Planning Board
P.O. Box 1230
435 Grand Avenue West
Chatham, Ontario N7M 5L8
Phone: (519) 631-1010

County of Elgin Planning Board
450 Sunset Drive
St. Thomas, Ontario N5R 5V1
Phone: (519) 631-1460

Regional Municipality of Haldimand-Norfolk
Planning Board
70 Towncentre Drive
Townsend, Ontario N0A 1S0
Phone: (416) 772-3571

Regional Municipality of Niagara
P.O. Box 1042
2201 St. David's Road
St. Thorold, Ontario L2V 4T7
Phone: (416) 685-1571

Check to be sure that the property in which you are interested is zoned for your intended use and that adjacent property zones do not conflict with your plans. For information, contact the city or county building inspector.

Coastal zone management

Contact the following agencies for questions concerning planning, resources, and development of the coastal zone.

Michigan Department of Natural Resources
Land Resources Programs Division
P.O. Box 30028
Lansing, MI 48909
Phone: (517) 373-1950

Ohio Department of Natural Resources
Division of Water
Fountain Square
Columbus, OH 43224
Phone: (614) 265-6717

Pennsylvania Department of Environmental Resources
Division Coastal Zone Management
P.O. Box 1467
Harrisburg, PA 17120
Phone: (717) 783-9500

New York Department of State
Coastal Management Program
162 Washington Avenue
Albany, NY 12231
Phone: (518) 474-3642

Environment Canada
Inland Waters Directorate
Ontario Region
Water Planning and Management Branch
P.O. Box 5050, 867 Lakeshore Road
Burlington, Ontario L7R 4A6
Phone: (416) 336-4712

Ontario Ministry of Natural Resources
Conservation Authorities and Water Management Branch
Whitney Block, Room 5620
99 Wellesley Street West
Toronto, Ontario M7A 1W3
Phone: (416) 965-6287

Consultants

It is inappropriate for the authors of this publication to endorse any individual or firm as a recommended coastal or construction consultant. We encourage prospective buyers as well as owners of existing property to seek expert advice on housing construction safety and site safety with respect to coastal hazards. The offices listed below under other topics and the offices of your local government are sources of advice on appropriate private consultants for your particular problem.

Disaster assistance

For information write or phone:

Michigan Department of State Police
111 South Capitol Avenue
Lansing, MI 48913
Phone: (517) 373-0617

Ohio Adjutant General's Department Office
2825 West Dublin-Granville Rd.
Worthington, OH 43085
Phone: (614) 889-7150

Pennsylvania Emergency Management Agency
Commonwealth Ave. and Forster St.
P.O. Box 3321
Harrisburg, PA 17120
Phone: (717) 783-8150

New York State Emergency Management Office
State Campus, Bldg. 22
Albany, NY 12226
Phone: (518) 457-6966

Environment Canada
Environment Protection Service
Ontario Region
25 St. Clair Avenue East 7th Floor
Toronto, Ontario M4T 1M2
Emergency (24 hr.) phone: (416) 973-5840

Transport Canada
Canadian Coast Guard
P.O. Box 2778
105 South Christina Street
Sarnia, Ontario N7T 7W1
Phone: (519) 337-6360
Emergency Operations Branch
Emergency (24 hr.) phone: (519) 337-6360

Search and Rescue Branch
Central Region Headquarters
1 Yonge Street 20th Floor
Toronto, Ontario M5E 1E5
Phone: (416) 973-1428

Ontario Ministry of Solicitor General
Emergency Planning
1st Floor 25 Grosvenor Street
Toronto, Ontario M7A 1Y6
Phone: (416) 965-6708

Ontario Ministry of Natural Resources
Conservation Authorities and Water Management Branch
Coordinator Emergency Management
Whitney Block Room 5637
99 Wellesley Street West
Toronto, Ontario M7A 1W3
Phone: (416) 965-6277

Dredging, filling, and construction in coastal waterways

Federal law requires that any person who wishes to dredge, fill, or place any structure in navigable water (almost any body of water in the United States) must apply for a permit from the U.S. Army Corps of Engineers. Information is also available on state requirements for this type of activity from:

Michigan Department of Natural Resources
P.O. Box 30028
Lansing, MI 48909
Phone: (517) 373-2329

Ohio Department of Natural Resources
Fountain Square
Columbus, OH 44324
Phone: (614) 265-6877

Pennsylvania Department of Environmental Resources
Bureau of Dams and Waterways Management
P.O. Box 1467
Harrisburg, PA 17120
Phone: (717) 787-2814

New York Department of Environmental Conservation
Region IX
600 Delaware Avenue
Buffalo, NY 14202
Phone: (716) 847-4560

U.S. Army Corps of Engineers
P.O. Box 1027
Detroit, MI 48231
Phone: (313) 226-6413

U.S. Army Corps of Engineers
1776 Niagara Street
Buffalo, NY 14207
Phone: (716) 876-5454

In Canada contact:

Public Works Canada
Ontario Regional Office
Regional Marine and Civil Engineering
4900 Yonge Street
Willowdale, Ontario M2N 6A6
Phone: (416) 224-4116

Fisheries and Oceans
Pacific and Freshwater Fisheries
Ontario Region
Small Craft Harbours
3050 Harvester Road
Burlington, Ontario L7N 3J1
Phone: (416) 336-6022

Public Works Canada
District Office
Southwestern Ontario
Marine and Civil Engineering
451 Talbot Street
P.O. Box 668, Station B
London, Ontario N6A 5C9
Phone: (519) 679-4298

Ontario Ministry of Natural Resources
Conservation Authorities and Water Management Branch
Whitney Block Room 5620
99 Wellesley Street West
Toronto, Ontario M7A 1W3
Phone: (416) 965-6287

Ontario Ministry of Natural Resources
Central Region
10670 Yonge Street
Richmond Hill, Ontario L4C 3C6
Phone: (416) 884-9203

Ontario Ministry of Natural Resources
Niagara District
Hwy. 20
P.O. Box 1070
Fonthill, Ontario L0S 1E0
Phone: (416) 892-2656

Niagara Peninsula Conservation Authority
Centre Street
Allanburg, Ontario L0S 1A0
Phone: (416) 227-1013

Grand River Conservation Authority
400 Clyde Road
Cambridge, Ontario N1R 5W6
Phone: (519) 621-2761

Ontario Ministry of Natural Resources
Southwestern Region
659 Exeter Road (Hwy 135)
London, Ontario N6A 4L6
Phone: (519) 681-5350

Ontario Ministry of Natural Resources
Simcoe District
P.O. Box 706
Hwy. 3
548 Queensway West
Simcoe, Ontario N3Y 4T2
Phone: (519) 426-7650

Long Point Region Conservation Authority
P.O. Box 525
Simcoe, Ontario N3Y 4N5
Phone: (519) 426-4623

Ontario Ministry of Natural Resources
Aylmer District
353 Talbot Street West
Aylmer, Ontario N5H 2S8
Phone: (519) 773-9241

Catfish Creek Conservation Authority
R.R. 5
Aylmer, Ontario N5H 2R4
Phone: (519) 773-9605

Kettle Creek Conservation Authority
R.R. 8
St. Thomas, Ontario N5P 3T3
Phone: (519) 631-1270

Ontario Ministry of Natural Resources
Chatham District
Kent County Municipal Building
435 Grand Avenue West
P.O. Box 1168
Chatham, Ontario N7M 5L8
Phone: (519) 354-7340

Lower Thames Valley Conservation Authority
100 Thames Street
Chatham, Ontario N7L 2Y8
Phone: (519) 335-3557

Essex Region Conservation Authority
860 Fairview Avenue West
Essex, Ontario N8M 1Y6

Environmental affairs

Administrator
Environmental Protection Agency
401 M Street, S.W.
Washington, DC 20460
Phone: (202) 755-2673

Michigan Department of Natural Resources
P.O. Box 30028
Lansing, MI 48909
Phone: (517) 373-1950

Ohio Environmental Protection Agency
361 East Broad Street
P.O. Box 1049
Columbus, OH 43216
Phone: (614) 466-8318

Pennsylvania Department of Environmental Resources
Fulton Bank Building
200 North Third Street
P.O. Box 2063
Harrisburg, PA 17120
Phone: (717) 783-3700

New York State Department of Conservation
50 Wolf Road
Albany, NY 12233
Phone: (518) 457-3446

Environment Canada
Ottawa, Ontario K1A 0H3
Phone: (613) 997-2800

Ontario Ministry of Environment
Suite 100
135 St. Clair Avenue West
Toronto, Ontario M4V 1P5
Phone: (416) 965-7117

Flood insurance

See *Insurance*.

Geologic information

Energy, Mines and Resources, Canada
Geological Survey of Canada
Scientific and Technical Enquiries
601 Booth Street
Ottawa, Ontario K1A 0E8
Phone: (613) 995-4089

Ontario Ministry of Natural Resources
Ontario Geological Survey
11th Floor 77 Grenville Street
Toronto, Ontario M7A 1W4
Phone: (416) 965-1283

Geologic Inquiries Group
U.S. Geological Survey
907 National Center
Reston, VA 22092
Phone: (703) 860-6517

Michigan Department of Natural Resources
Geological Survey Division
Stevens T. Mason Building
P.O. Box 30028
Lansing, MI 48909
Phone: (517) 334-6923

Ohio Department of Natural Resources
Division of Geological Survey
Fountain Square
Columbus, OH 43224
Phone: (614) 265-6605

Pennsylvania Department of Environmental Resources
Bureau Topographic & Geologic Survey
P.O. Box 1467
Harrisburg, PA 17120
Phone: (717) 787-2169

New York State Geological Survey
Office of Cultural Education
Empire State Plaza
Albany, NY 12230
Phone: (518) 474-5816

Hazards

See also *Shore erosion and insurance.*

Office of Ocean and Coastal Resources Management
National Oceanic and Atmospheric Administration
3300 Whitehaven Street, N.W.
Washington, DC 20235

Health

See also *Sanitation and septic system permits.*

The local health department in your city or county will provide information on home waste-treatment systems, water-supply systems, and similar health matters. Questions relating to water quality and health matters also may be directed to:

Environment Canada
Inland Waters Directorate, Ontario Region
Water Quality Branch (Ontario Region)
P.O. Box 5050, 867 Lakeshore Road
Burlington, Ontario L7R 4A6
Phone: (416) 336–4662

Ontario Ministry of the Environment
Laboratory Services Branch
Water Quality Section
P.O. Box 213 Resources Road
Rexdale, Ontario M9W 5L1
Phone: (416) 248–3512

Michigan Department of Public Health
3500 North Logan Street
Baker-Olin West Building
P.O. Box 30035
Lansing, MI 48909
Phone: (517) 373–1320

Ohio Department of Health
246 North High Street
Columbus, OH 43215
Phone: (614) 466–2253

Erie County Health Department
606 West 2nd Street
Erie, PA 16507
Phone: (814) 454–5811

New York Department of Health
Tower Bldg. Empire State Plaza
Albany, NY12237
Phone: (518) 474–2011

History

See *Archives and records.*

Insurance

In coastal areas special building requirements must often be met in order to obtain windstorm insurance or affordable flood insurance. To find out the requirements for your area, check with your insurance agent. To determine eligibility for the U.S. National Flood Insurance Program contact your coordinating office or the FEMA office for your region.

Water Management Division
Michigan Department of Natural Resources
P.O. Box 30028
Lansing, MI 48909
Phone: (517) 373-3930

Ohio Department of Natural Resources
Flood Plain Planning Unit
Fountain Square
Columbus, OH 43224
Phone: (614) 265-6755

Pennsylvania Department of Community Affairs
551 Forum Building
Harrisburg, PA 17120
Phone: (717) 787-7400

Flood Protection Bureau
New York Department of Environmental Conservation
50 Wolf Road, Room 422
Albany, NY 12233
Phone: (518) 457-3157

FEMA Region V Office
300 South Wacker Drive
24th Floor
Chicago, IL 60606
Phone: (312) 353-1500
(Michigan and Ohio)

FEMA Region III Office
Curtis Building
Sixth and Walnut Streets
Philadelphia, PA 19106
Phone: (215) 597-9416
(Pennsylvania)

FEMA Region II Office
26 Federal Plaza
Room 1349
New York, NY 10007
Phone: (212) 264-4756
(New York)

Your insurance agent or community building inspector should
be able to provide you with information about the location of
your building site on the flood insurance rate map (FIRM), and
the elevation required for the first floor to be above the 100-year
flood level. If they cannot provide this information, request the
FIRM for your area from the above state or FEMA regional address.
Note that a flood policy under the national flood insurance program is separate from your regular home owner's policy. General
information on the National Flood Insurance Program is also
available from:

National Flood Insurance Program
P.O. Box 619
Lanham, MD 20706
Phone: 1-800-638-6620

For other types of insurance questions contact:

Insurance Bureau
Michigan Department of Licensing and Regulation
1048 Pierport Street
P.O. Box 30220
Lansing, MI 48909
Phone: (517) 373-0220

Ohio Department of Insurance
2100 Stella Court
Columbus, OH 43215
Phone: (614) 466-2691

Pennsylvania Insurance Department
1326 Strawberry Square
4th and Walnut Street
Harrisburg, PA 17120
Phone: (717) 787-5288

New York Insurance Department
Two World Trade Center
New York, NY 10047
Phone: (212) 488-4124

Lake levels

U.S. Army Corps of Engineers
Detroit District
P.O. Box 1027
Detroit, MI 48231
Phone: (313) 226-6413
(request the "Monthly Bulletin of Lake Levels for the Great Lakes")

Great Lakes Environmental Research Laboratory
National Ocean Service, NOAA
2300 Washtenaw Avenue
Ann Arbor, MI 48104
Phone: (313) 374-7100

Environment Canada
Great Lakes Water Level Communications Centre
P.O. Box 5050
867 Lakeshore Road
Burlington, Ontario L7R 4A6
Phone: (416) 336-4581

Land acquisition

When acquiring property or a condominium—whether in a subdivision or not—consider the following: (1) The description and survey of land in coastal areas can be very complicated. Old titles granting fee-simple rights to property below lake level may not be upheld in court; titles should be reviewed by a competent attorney before they are transferred. (2) Ask about the provision of sewage disposal and utilities including water, electricity, gas, and telephone. (3) Be sure any promises of future improvements, access, utilities, additions, common property rights, etc., are in writing. (4) Be sure to visit the property and inspect it carefully before buying it. For specific information contact:

Environment Canada
Inland Waters Directorate
Water Planning and Management Branch
P.O. Box 5050
867 Lakeshore Road
Burlington, Ontario L7R 4A6
Phone: (416) 336–4712

Ontario Ministry of Natural Resources
Land Management Branch
Public Land Section Land Acquisition
Whitney Block Room 6629
99 Wellesley Street West
Toronto, Ontario M7A 1W3
Phone: (416) 965–2743

For information on past, present, and future lake levels contact:

Michigan Department of Licenses and Registration
808 Southland Avenue
Lansing, MI 48909
Phone: (517) 373–6338

Ohio Department of Natural Resources
Division of Water
1939 Fountain Square Court
Columbus, OH 43224
Phone: (614) 265–6717

Pennsylvania Department of General Services
Commonwealth Avenue & North Street
Harrisburg, PA 17125
Phone: (717) 787–2834

New York Department of State
162 Washington Avenue
Albany, NY 12231
Phone: (518) 474–4429

Maps

A wide variety of maps are useful to planners and managers and may be of interest to individual property owners. Topographic, geologic, and land-use maps and orthophoto quadrangles are available from:

Energy, Mines and Resources, Canada
Surveys and Mapping Branch
25 St. Clair Avenue East, 9th Floor
Toronto, Ontario M4T 1M2
Phone: (416) 973-7503

Canada Map Office
130 Bentley Avenue
Ottawa, Ontario
Phone: (613) 998-9900

Ontario Ministry of Natural Resources
Public Information Centre
Maps Info Distribution
Whitney Block Room 1640
99 Wellesley Street West
Toronto, Ontario M7A 1W3
Phone: (416) 965-6511

Distribution Section
U.S. Geological Survey
Federal Center Building 41
Box 25286
Denver, CO 80225
Phone: (303) 234-3832

A free index to the type of map desired (for example, the "Index to Topographic Maps of Pennsylvania") should be requested and then used for ordering specific maps.

For evacuation maps, call your county department of Emergency Preparedness. For flood-zone maps, see *Insurance*. For planning maps, call or write your local county commission. For soil maps and septic suitability, see *Soils*.

Nautical charts in several scales contain navigation information for Lake Erie. Nautical-chart index maps are available from:

Fisheries and Oceans
Bayfeld Laboratory for Marine Science and Surveys
Marine Information Centre
P.O. Box 5050
867 Lakeshore Boulevard
Burlington, Ontario L7R 4A6
Phone: (416) 336-4843

The Book Store
Windsor Public Library
850 Ouellette Avenue
Windsor, Ontario N9A 4M9
Phone: (519) 225-6765

Oxford Book Shops, Ltd.
742 Richmond Street
London, Ontario N6A 1L6
Phone: (519) 438-8336

Bell Marine and Mill Supply LTD.
West Pier Road Box 398
Port Colborne, Ontario L3K 1B7
Phone: (416) 835-1144

National Ocean Survey
Distribution Division (C-44)
National Oceanic and Atmospheric Administration
Riverdale, MD 20840
Phone: (301) 436-6990

National Wetland Inventory (NWI) maps for areas mapped in the state of Pennsylvania can be acquired by writing or calling the:

Pennsylvania Department of Environmental Resources
Division of Coastal Zone Management
P.O. Box 1467
Harrisburg, PA 17120
Phone: (717) 783-9500

All other requests for wetland mapping, other than for the state of Pennsylvania, should be made to the:

U.S. Fish and Wildlife Service
One Gateway Center
Suite 700
Newton Corner, MA 02158
Phone: (617) 965-5100, ext. 379

Movies and audiovisual materials

For information contact the following state Sea Grant offices:

Michigan Sea Grant College Program
The University of Michigan
2200 Bonisteel Blvd.
Ann Arbor, MI 48109
Phone: (313) 763-1437

Ohio Sea Grant Program
Ohio State University
484 West 12th Avenue
Columbus, OH 43210
Phone: (614) 422-8949

Pennsylvania Department of Environmental Resources
Division of Coastal Zone Management
P.O. Box 1467
Harrisburg, PA 17120
Phone: (717) 783-9500

New York Sea Grant Institute
37 Elk Street
Albany, NY 12246
Phone: (518) 436-0701

Fisheries and Oceans
Communications Directorate
Creative Services Division
240 Sparks Street
Ottawa, Ontario K1A 0E6
Phone: (613) 993-0600

Environment Canada
Inland Waters Directorate, Ontario Region
Water Planning and Management Branch
P.O. Box 5050
867 Lakeshore Road
Burlington, Ontario L7R 4A6
Phone: (416) 336-4712

National Film Board
Distribution Branch
1 Lombard Street
Toronto, Ontario M5C 1J6
Phone: (416) 973-2235

Parks and recreation

Coastal state and provincial parks are scattered along the Lake Erie shore. For information on parks contact the Ontario Ministry of Natural Resources and the following state agencies:

Parks Division
Michigan Department of Natural Resources
P.O. Box 30028
Lansing, MI 48909
Phone: (517) 373-1270

Division of Parks and Recreation
Ohio Department of Natural Resources
Fountain Square, Building C
Columbus, OH 43224
Phone: (614) 265-6511

Bureau of State Parks
Pennsylvania Department of Environmental Resources
P.O. Box 1467
Harrisburg, PA 17120
Phone: (717) 787-6640

New York Office of Parks, Recreation, and Historic Preservation
Agency Building #1
Empire State Plaza
Albany, NY 12238
Phone: (518) 474-0443

Planning and land use

See also *Coastal zone management.*

For specific information on your area, check with the local town or county commission. Most local governments have planning boards that answer to the commission and have available copies of existing or proposed land-use plans.

Roads and property access

Rules and regulations vary from state to state and province to province. Before buying, determine if access rights and roads will be provided. If connecting a driveway to a state-maintained right-of-way, you will probably need a permit from the highway department. Contact a state, provincial, or county road official.

Michigan Department of Transportation
425 West Ottawa Street
P.O. Box 30050
Lansing, MI 48909
Phone: (517) 373-2114

Ohio Department of Transportation
25 South Front Street
Columbus, OH 43215
Phone: (614) 466-2335

Pennsylvania Department of Transportation
1200 Transportation and Safety Building
Harrisburg, PA 17120
Phone: (717) 787-5574

New York Department of Transportation
State Campus, Bldg. 5
Albany, NY 12232
Phone: (518) 457-4422

In Canada:

Transport Canada
Canadian Surface Transportation Administration
Surface Policy Planning and Urban Programs
Place de Ville Tower C
Transport Canada Building
Queen and Lyon Streets
Ottawa, Ontario K1A 0N5
Phone: (613) 995-2326

Public Works
Ontario Region Office
Property Administration
Property Management
4900 Yonge Street
Willowdale, Ontario M2N 6A6
Phone: (416) 224-4276

Transportation and Communication
Highway Engineering Division
Environmental Office
Surveys and Plans Office
Lower Floor East Building
1201 Wilson Avenue
Downsview, Ontario M3M 1J8
Phone: (416) 248-3627

Sanitation and septic system permits

See also *Water Resources*.

Where property has no access to a sewer system, it is usually necessary to obtain a permit for a septic system from the local health department *before* a construction permit can be issued. Such a permit is issued only if the soil is suitable for a septic system. Clay-rich soils are usually unsuitable. Likewise, if your property does not have access to a municipal water system you will need a well. Check with the county health department to determine the quality of the local groundwater. Make sure that the design and location of your septic system will safeguard your water supply.

Activity resulting in effluent discharge or runoff into surface waters requires certification from the state water pollution control agency that the proposed activity will not violate water quality standards. For information contact:

Michigan Department of Public Health
3500 North Logan Street
P.O. Box 30035
Lansing, MI 48909
Phone: (517) 373-1320

Ohio Department of Health
246 North High Street
Columbus, OH 43215
Phone: (614) 466-2253

Erie County Health Department
606 West 2nd Street
Erie, PA 16507
Phone: (814) 454-5811

New York Department of Environmental Conservation
Region IX
600 Delaware Avenue
Buffalo, NY 14202
Phone: (716) 847-4560

In Ontario contact the municipal offices listed under *Building codes and zoning*, and:

Ontario Ministry of the Environment
Water Resources Branch
Water and Wastewater Management Section
7th Floor
1 St. Clair Avenue West
Toronto, Ontario M4V 1K6
Phone: (416) 965-1655

A permit for the construction of a sewage disposal structure or any other structure in navigable waters of the United States must be obtained from the U.S. Army Corps of Engineers.

A permit for any discharge into navigable waters of the United States must be obtained from the U.S. Environmental Protection Agency. Recent judicial interpretation of the Federal Water

Pollution Control Amendments of 1972 extends federal jurisdiction for protection of wetlands above the mean high-water mark. Federal permits may now be required for the development of land that occasionally is flooded by water draining indirectly into a navigable waterway. Information may be obtained from:

Enforcement Division
Environmental Protection Agency
Region V
230 South Dearborn Street
Chicago, IL 60604
Phone: (312) 353-5250
(Michigan and Ohio)

Enforcement Division
Environmental Protection Agency
Region III
6th and Walnut Streets
Philadelphia, PA 19106
Phone: (215) 597-3642
(Pennsylvania)

Enforcement Division
Environmental Protection Agency
Region II
26 Federal Plaza
New York, NY 10278
Phone: (212) 264-2525
(New York)

Environment Canada
Environmental Protection Service
Ontario Region
Pollution Control Division
25 St. Clair Avenue East 7th Floor
Toronto, Ontario M4T 1M2
Phone: (416) 973-5840
(Canada)

Ontario Ministry of Health
Public Health Branch
London Regional Office
3rd Floor
227 Queen's Avenue
London, Ontario N6A 1J8
Phone: (519) 432-1866
(Ontario)

Shore erosion

Information on shore erosion, flooding, and lake levels is available from the different federal, provincial, and state agencies listed under *Dredging, filling, and construction* in coastal waterways. In Canada information is also available from:

Environment Canada
Great Lakes Water Level Communication Centre
P.O. Box 5050
867 Lakeshore Road
Burlington, Ontario L7R 4A6
Phone: (416) 336-4581

Soils

See also *Sanitation and vegetation.*

Soil type is important in terms of: the type of vegetation it can support; the type of construction technique it can withstand (for example, loading, support of piling); its drainage characteristics; and its ability to accommodate septic systems. The following agencies cooperate to produce a variety of maps and reports useful to property owners:

Agriculture Canada
Land Resource Research Institute
Soil Inventory Section
Sir John Carling Building
930 Carling Avenue
Ottawa, Ontario K1A 0C5
Phone: (613) 995-5011

Environment Canada
Lands Directorate
P.O. Box 5050
867 Lakeshore Road
Burlington, Ontario L7R 4A6
Phone: (416) 336-4551

Ontario Ministry of Agriculture and Food
Agricultural Representatives Branch
Guelph Agriculture Centre
P.O. Box 1030
Guelph, Ontario N1H 6N1
Phone: (519) 823-5700

U.S. Department of Agriculture
Soil Conservation Service
P.O. Box 2890
Washington, DC 20013
Phone: (202) 447-4543

Michigan Soil Conservation Service
1405 South Harrison Road, Room 101
East Lansing, MI 48823
Phone: (517) 337-6702

Ohio Soil Conservation Service
200 North High Street
Room 522
Columbus, OH 43215
Phone: (614) 469-6962

Erie County Conservation District
R.D. 5, Route 19
Waterford, PA 18441
Phone: (814) 796-4203

New York Soil Conservation Service
J. M. Hahley Federal Building, Room 771
100 South Clinton Street
Syracuse, NY 13260
Phone: (315) 423-5521

Your community or county health department usually can provide soils information relative to construction and septic permits, or refer you to another agency for specific soil information.

Subdivisions

U.S. subdivisions containing more than 100 lots and offered in interstate commerce must be registered with the Office of Interstate Land Sales Registration (as specified by the Interstate Land Sales Full Disclosure Act). Prospective buyers must be provided with a property report. This office also produces a booklet entitled *Get the Facts before Buying Land* for people who wish to invest in land. Information on subdivision property and land investment is available from:

Office of Interstate Land Sales Registration
U.S. Department of Housing and Urban Development
Washington, DC 20410
Phone: (202) 755-6600

In Canada:

Ontario Ministry of Natural Resources
Land Management Branch
Public Lands Section
Room 6603
99 Wellesley Street West
Toronto, Ontario M7A 1W3
Phone: (416) 965-2743

For information on real estate in your state contact:

Realty and Environmental Services Bureau
Michigan Department of Licensing and Regulation
808 Southland Avenue
Lansing, MI 48909
Phone: (517) 373-6338

Division of Real Estate
Ohio Department of Commerce
Two Nationwide Plaza
Chestnut and High Streets
Columbus, OH 43215
Phone: (614) 466-4100

Bureau of Real Estate
Pennsylvania Deprtment of General Services
503 North Office Building
Commonwealth Avenue and North Street
Harrisburg, PA 17125
Phone: (717) 787-2834

Division of Licensing Services
New York Department of State
162 Washington Avenue
Albany, NY 12231
Phone: (518) 474-4429

Vegetation

Information on vegetation may be obtained from the local Soil and Water Conservation District.

Water resources

A variety of agencies are concerned with water quality; availability; permit issuance and construction of wells; water contamination; permit issuance, regulation, and construction of waste-disposal systems; and sites. Many agencies, including Environment Canada, the U.S. Army Corps of Engineers district offices, and the different agencies for natural and environmental resources, can provide information. Agencies include:

Water Quality Division
Michigan Department of Natural Resources
Stevens T. Mason Building
P.O. Box 30028
Lansing, MI 48909
Phone: (517) 373-1947

Office of Wastewater Pollution
Ohio Environmental Protection Agency
Seneca Towers
361 East Broad Street
Columbus, OH 43215
Phone: (614) 466-7427

Bureau of Water Quality Management
Pennsylvania Department of Environmental Resources
Fulton Bank Building
200 North 3rd Street
P.O. Box 2063
Harrisburg, PA 17120
Phone: (717) 787-2666

Division of Water
New York Department of Environmental Conservation
50 Wolf Road
Albany, NY 12233
Phone: (518) 457-6674

Environment Canada
Inland Waters Directorate, Ontario Region
Water Quality Branch
P.O. Box 5050
867 Lakeshore Road
Burlington, Ontario L7R 4A6
Phone: (416) 336–4663

Ontario Ministry of Environment
Water Resources Branch
Water and Wastewater Management Section
7th Floor
1 St. Clair Avenue West
Toronto, Ontario M4V 1K6
Phone: (416) 965–1655

Weather

Environment Canada
Great Lakes Water Level Forecast Centre
Ontario Weather Centre
Box 159 (AMF)
Toronto, Ontario L5P 1B1
Phone: (416) 676–3019

Environment Canada
Toronto Weather Office
Lester P. Pearson International Airport
Atmospheric Environment Service
4905 Dufferin Street
Downsview, Ontario M3H 5T4
Weather Phone: (416) 676–3066
Marine Phone: (416) 676–4567

General information may be obtained from:

National Weather Service (NOAA) Eastern Region
585 Stewart Avenue
Garden City, NY 11530
Phone: (516) 248–2101

National Weather Service (NOAA) Central Region
601 East 12th Street
Kansas City, MO 64106
Phone: (816) 374–5463

In Canada, the Environment Canada Toronto forecast office issues lakeshore flooding and erosion forecasts for the Canadian shore of Lake Erie, and in the U.S., the NOAA National Weather Service offices at Ann Arbor (the Michigan shore), Cleveland (the Ohio and Pennsylvanian shores), and Buffalo (the New York shore) issue the flooding and erosion forecasts. "Statements" are issued when the lake level (determined by water level gage records) reaches 5 feet above low water datum (573.6 ft IGLD) but is not expected to exceed 6 feet above LWD. "Warnings" are issued if lake level is expected to exceed 6 feet (personal communication, Deron Boyce, U.S. (National Weather Service).

Wildlife

Contact the department of natural resources or environmental resources in each state except for Pennsylvania. In Pennsylvania contact:

Pennsylvania Fish Commission
3532 Walnut Street (Progress)
P.O. Box 1673
Harrisburg, PA 17105
Phone: (717) 657-4519

Pennsylvania Game Commission
8000 Derry Street (Rutherford)
P.O. Box 1567
Harrisburg, PA 17105
Phone: (717) 787-4250

In Canada contact:

Environment Canada
Canadian Wildlife Service
P.O. Box 5050
867 Lakeshore Road
Burlington, Ontario L7R 4A6
Phone: (416) 336-4960

Environment Canada
Canadian Wildlife Service
152 Newbold Court
London, Ontario N6E 1Z7
Phone: (519) 681-0486

Zoning

See *Building codes and zoning.*

Appendix B
Useful references

The following publications are listed by subject; they are arranged in the approximate order that they appear in this book. A brief description of most of the references is provided, and sources are included for those readers who would like more information on a particular subject. Many of the references distributed by government agencies listed are either low in cost or free. We encourage the reader to take advantage of these informative publications.

History

1. *Lake Erie*, by Harlan Hatcher, 1945. This classic account traces the history of the lake from discovery by Europeans through World War II. Published by the Bobbs-Merrill Company, NY.

2. *Erie—the Lake that Survived*, by Noel M. Burns, 1985, is an excellent review of the lake as a chemical and biological system. Written for the layman, the text is technical only to the degree necessary to discuss chemical contamination, nutrient loading, biological communities, and ecosystems. Geological development and human history are discussed as controls on the lake system, as are lake processes. This text should be of interest to all citizens of the Lake Erie drainage basin and students of limnology. Although not by intent, this reference and the current volume cover complementary aspects of Lake Erie. The 320-page book, published by Rowman and Allanheld, Totowa, NJ 07512, should be available through your bookstore and library.

3. *The Late, Great Lakes*, by William Ashworth, 1986. This work is recommended reading for all Great Lakes citizens, especially anyone who has grown complacent, thinking the lakes' problems have been solved. The author does not hold back, and in a "tell it like it is" style, he reviews how the lakes region was settled and developed, the problems that emerged, and what we may think has been solved. Pollution is a central theme, but all aspects of the lakes are treated, including the erosion problem. This 274-page book should receive a wide reading—and we hope it will make an impact on public attitude. Published by Alfred A. Knopf, Inc., NY, and available through your bookstore or library.

4. *Man and the Marine Environment*, edited by Robert A. Ragotzkie, 1983, is a collection of essays that are relevant to the Great Lakes as well as to the oceans. Great Lakes readers will be particularly interested in the chapters entitled "Shipping in the Great Lakes" (3), "Coastal Management: An Unfinished Undertaking" (7), and "The Great Lakes: A Microcosm of the World Ocean" (9). Published by CRC Press, Inc., Boca Raton, FL 33431 and available through your bookstore or library.

5. *The Mapping of Ohio*, by T. H. Smith, 1977. Ohio's history

is traced through a compilation and review of historic maps beginning in the mid-17th century and concluding with the railroad map of 1875. This 252-page book is of interest primarily to students of history, but it sheds some light on the early understanding of Lake Erie. Published by Kent State University Press, Kent, OH 44242.

6. *Ohio Geological Survey 2nd Annual Report*, Geological Report by Charles Whittlesey, pp. 41–71, 1838. The first documentation of erosion rates on Lake Erie is recorded in this out-of-print report.

7. *Lake Erie Floods, Lake Levels, Northeast Storms, and the Formation of Sandusky Bay and Cedar Point*, by Edwin L. Moseley, 1904, presents early observations and hypotheses by one of the first prominent Lake Erie scientists. Reprinted in 1973 by the Ohio Historical Society, Columbus, OH.

8. *Engineers for the Public Good: A History of the Buffalo District U.S. Army Corps of Engineers*, by Nuala Drescher, 1982. This comprehensive work includes chapters that range from early exploration and development of the Lake Erie and Lake Ontario region to present-day Corps of Engineers projects and planning. Published by U.S. Army Corps of Engineers, Buffalo, NY.

Ecology, recreation, and wildlife

9. *The Aging Great Lakes*, by C. F. Powers and A. Robertson, 1966. The authors discuss the process of eutrophication, using Lake Erie as an example of accelerated eutrophication. Published in *Scientific American*, vol. 215, no. 5, pp. 95–104.

10. *Changes in the Environment and Biota of the Great Lakes*, by A. M. Beeton, 1969. A general summary by one of the foremost biological scientists on the Great Lakes: In *Eutrophication: Causes, Consequences, Correctives*, Printing and Publishing Office, National Academy of Sciences, Washington, DC, pp. 150–87.

11. *Lake Erie: Effects of Exploitation, Environmental Changes, and New Species on the Fishery Resources*, by W. L. Hartman, 1972. The effects of man in terms of exploitation, environmental changes, and new fish species on Lake Erie's fish. Published in the *Journal of the Fisheries Research Board of Canada*, vol. 29, no. 6, pp. 899–912.

12. *Guide to Fishing Reefs in Western Lake Erie*, by C. E. Herdendorf, 1980. A practical guide that includes bathymetric maps of the reefs and preferred sport-fish habitats. Available from Ohio Sea Grant, Columbus, OH 43210.

13. *Guide to Fishing in Central Lake Erie*, by C. E. Herdendorf, 1984. Includes useful information on the physical and biological nature of central Lake Erie. Available from Ohio Sea Grant, Columbus, OH 43210.

14. *Recreational Fishing in Central Lake Erie*, by C. E. Herdendorf, F. R. LichtKoppler, D. O. Kelch, and F. L. Snyder, 1984. This guide provides basic data on where to fish as well as tips on how to fish in Lake Erie. Available from Ohio Sea Grant, Columbus, OH 43210.

15. *Ohio's Natural Heritage*, edited by Michael B. Lafferty, 1979. Published by the Ohio Academy of Science, 445 King Ave., Columbus, OH 43210, 324 pp. In addition to the excellent general description of the natural environments of Ohio, this book includes chapters dealing with Lake Erie (13), geologic history (2, 3), water (8), and land use (17). Recommended to all residents, and to visitors who enjoy understanding nature in the area of their travels. Available through your library and local bookstore.

Geography and geology

16. *The Physiography of Southern Ontario*, by L. J. Chapman and D. F. Putman, 1966. The classic reference on physical geography for the Canadian portion of the Lake Erie basin. Published by the University of Toronto Press, Toronto, Ontario.

17. *Geology of the Great Lakes*, by Jack L. Hough, 1958. This classic review of Great Lakes geology emphasizes the origin and Pleistocene history of the lakes. Published by the University of Illinois Press, Urbana, IL.

18. *Pleistocene Deposits of the Erie Lobe*, by R. P. Goldthwait, Aleksis Dreimanis, J. L. Forsythe, P. F. Karrow, and G. W. White, 1965. A summary of the glacial deposits in the Lake Erie area written by experts from Canada and the United States. In *The Quaternary of the United States*, edited by Herbert E. Wright and David G. Frey. Published by Princeton University Press, Princeton, NJ.

19. *Glacial Geology of Northeastern Ohio*, Ohio Geological Survey Bulletin 68, by George W. White, 1982. A good summary of White's glacial geologic work along the south shore of Lake Erie is presented. Stanley M. Totten wrote the chapter on Pleistocene beaches and strandlines. Published by the Ohio Geological Survey, Columbus, Ohio 43224.

20. *The Geologic Interpretation of Scenic Features in Ohio*, by J. Ernest Carman, 1946. A general article in the *Ohio Journal of Science*, vol. 46, pp. 241–83, that includes a section on the Lake Erie basin and islands.

21. *Coastal Geomorphology and Geology of the Ohio Shore of Lake Erie*, by C. H. Carter, D. E. Guy, Jr., and J. A. Fuller, 1981. This is a field trip guide prepared for the Geological Society of America, Cincinnati meeting. *Geological Society of America Guidebook*, vol. 3, pp. 433–56.

22. *Nearshore Sediment Studies in Western Lake Erie*, by J. P. Coakley, 1972. The nature and distribution of bottom sediments are mapped from Wheatley to the Detroit River. The source of the sediment is also discussed. Published by the International Association Great Lakes Research, proceedings of the 15th conference, Toronto, Ontario, pp. 330–43.

23. *Lake Erie Nearshore Sediments — Mohawk Point to Port Burwell, Ontario*, by D. A. St. Jacques and N. A. Rukavina, 1973. The authors have mapped the nature and distribution of sediment and sedimentary rocks in the shore and nearshore zones. Published by the International Association Great Lakes Research, proceedings of the 16th conference, Toronto, Ontario,

pp. 454–67.

24. *Lake Erie Nearshore Sediments—Fort Erie to Mohawk Point, Ontario*, by N. A. Rukavina and D. A. St. Jacques, 1971. The authors have mapped the nature and distribution of sediment and sedimentary rocks in the nearshore zone. Published by the International Association Great Lakes Research, proceedings of the 14th conference, Toronto, Ontario, pp. 387–93.

25. *Physical characteristics of the Reef Area of Western Lake Erie*, by C. E. Herdendorf and L. L. Braidech, 1972. The report includes research on modern lake processes as well as the geology of the island area. Published by the Ohio Geological Survey, Report Investigations 82, Columbus, OH 43224.

26. *Sand Resources of Southern Lake Erie, Conneaut to Toledo, Ohio—A Seismic Reflection and Vibracore Study*, by S. J. Williams and others, 1980. The survey documents two off-shore areas, Fairport Harbor and Lorain-Vermilion, as having good potential for beach restoration projects. Miscellaneous Report 80-10, U.S. Army Corps of Engineers, Coastal Engineering Research Center, Vicksburg, MS 39180.

27. *Regional Geology of the southern Lake Erie (Ohio) Bottom: A Seismic Reflection and Vibracore Study*, by C. H. Carter and others, 1982. A comprehensive survey that documents the nature, distribution, and geometry of deposits within 15 kilometers of the Ohio shore. Miscellaneous Report 82-15, U.S. Army Corps of Engineers, Coastal Engineering Research Center, Vicksburg, MS 39180.

28. *Geological Character and Mineral Resources of South Central Lake Erie*, by S. Jeffress Williams and Edward P. Meisburger, 1982. A technical report on the area between the Ohio-Pennsylvania border and Erie, Pennsylvania, evaluating the potential for recovery of sand and gravel for beach nourishment at Presque Isle. Miscellaneous Report 82-9, U.S. Army Corps of Engineers, Coastal Engineering Research Center, Vicksburg, MS 39180.

29. *Southern Great Lakes, Field Guide*, by R. M. Feldmann and others, 1975, includes a chapter on coastal geology. The guide provides a good field introduction to the Ohio portion of the Lake Erie shoreline. Published as part of a national geology guidebook series by Kendall-Hunt Publishing Company, Dubuque, IA 52001, and available through college libraries.

30. *Geoarchaeological Interpretation of Great Lakes Coastal Environments*, by Curtis E. Larsen, 1985. Although a technical paper, this publication will appeal to those readers interested in the relationship of early changing lake levels and native American cultures. Perhaps the most interesting aspect of the paper is the suggestion that the modern Great Lakes have been significantly higher in the last 2,000 years than they are at present. Published in *Archaeological Sediments in Context*, edited by J. K. Stein and W. R. Farrand, pp. 91–110, Center for the Study of Early Man, Institute of Quaternary Studies, University of Maine, Orono, ME 14473, and available through college and university libraries.

31. *An Introduction to Coastal Geomorphology*, by John Pethick, 1984. This is a good introduction to some of the com-

plex factors that influence the shape of the coastline. Recommended as a starter for anyone who is interested in gaining a better understanding of coastal systems. Published by Edward Arnold, 300 North Charles Street, Baltimore, MD 21201.

32. *Waves and Beaches*, by Willard Bascom, 1980. An easily understandable discussion of beaches and coastal processes. Published by Anchor Books-Doubleday, Garden City, NY 11530. Available in paperback from local bookstores.

33. *Beaches and Coasts*, 2nd edition, by C. A. M. King, 1972. Classic, technical treatment of beach and coastal processes. Published by St. Martin's Press, 175 Fifth Avenue, New York, NY 10010.

34. *Beach Processes and Sedimentation*, by Paul Komar, 1976. The most up-to-date technical explanation of beaches and beach processes. Recommended only to serious students of the beach. Published by Prentice-Hall, Englewood Cliffs, NJ 07632.

35. *Our Changing Coastlines*, by Francis Shepard and Harold Wanless, 1971. A classic text that describes the coastal features of the United States. Published by McGraw-Hill, New York, NY.

36. *Encyclopedia of Beaches and Coastal Environments*, by Maurice Schwartz, 1982. A comprehensive survey of the field of coastal studies, including geomorphic, biologic, engineering, and human aspects of the world's coasts in an easy-to-use encyclopedia format. Good as a general reference. Published by Dowden, Hutchinson, and Ross, Stroudsburg, PA, and available in university libraries.

37. *Coastal Mapping Handbook*, edited by M. Y. Ellis, 1978.

A primer on coastal mapping outlining the various types of maps, charts, and photographs available; sources for such products; data and uses; and state coastal mapping programs. Includes information appendixes and examples. A valuable starting reference for anyone interested in maps or mapping. For sale by the Superintendent of Documents, U.S. Government Printing Office, Washington, DC 20402 (stock no. 024-001-03046-2), 200 pp.

Lake levels, storms, waves, weather

38. *Monthly Water Level Bulletin*, by Department of Fisheries and Oceans, and Environment Canada. Similar to the monthly bulletin published by the U.S. Army Corps of Engineers. Copies available from Environment Canada, Burlington, Ontario.

39. *Great Lakes Water Levels*, by Peter Yee and Jim Lloyd of Environment Canada, 1985. A succinct (20-page) booklet covering topics such as factors affecting water levels to forecasts of water levels. Available from Environment Canada, Burlington, Ontario.

40. *Monthly Bulletin of Lake Levels for the Great Lakes*, by the Detroit District, U.S. Army Corps of Engineers. The bulletin provides recorded levels for the previous year and the current year as well as the forecast levels for the following six months. Lakeshore property owners should request that their names be placed on the mailing list to receive the bulletins. Published by the U.S. Army Corps of Engineers, P.O. Box 1027, Detroit, MI

48231.

41. *Great Lakes Water Level Facts*, by the U.S. Army Corps of Engineers, Detroit District, 1985. A helpful 15-page booklet outlining the Great Lakes. Free from the Detroit District, U.S. Army Corps of Engineers, P.O. Box 1027, Detroit, MI 48231.

42. *A Forecast Model for Great Lakes Water Levels*, by Barry P. Cohn and Joseph E. Robinson, 1976. Spectral analysis is used to define periodic cycles in lake levels. Published in *Journal of Geology*, vol. 84, pp. 455–65.

43. *November 1972 Floods on the Lower Great Lakes*, by Anthony J. Brazel and David W. Phillips, 1974. The paper summarizes the 1972 storm and its effects both on Lake Erie and on Lake St. Clair. Published in *Weatherwise*, pp. 56–62.

44. *The November 1972 Storm on Lake Erie*, by Charles H. Carter, 1973. Documentation of the storm and storm effects is presented in this report published by the Ohio Geological Survey, Fountain Square, Bldg. B, Columbus, OH 43224.

45. *Great Lakes Storm Surge of April 6, 1979*, by P. F. Hamblin, 1979. Documentation of the wind tide as well as comparison of a hindcast of the wind tide is presented. Published in the *Journal of Great Lakes Research*, vol. 5, pp. 312–15.

46. *Wave and Lake Level Statistics for Lake Erie*, by Thorndike Saville, Jr., 1953. The classic reference on waves and lake level; discussion focuses on four sites along the U.S. shore: Monroe, MI, Cleveland, OH, Erie, PA, and Buffalo, NY. Published by the U.S. Army Corps of Engineers as Beach Erosion Board Technical Memorandum 37, Washington, DC.

47. *The Climate of the Great Lakes Basin*, by D. W. Phillips and J. A. W. McCulloch, 1972. A thorough report on the overall weather and climate. Published by Department of Environment, Climatological Studies 20, Ottawa, Ontario.

48. *Synthesized Winds and Wave Heights for the Great Lakes*, by T. L. Richards and D. W. Phillips, 1970. A report similar in a general way to Saville's work on the U.S. shore (reference 46). Published by Department of Transport, Meteorological Branch, Climatological Studies 17, Ottawa, Ontario.

49. *Climatology and Weather Services of the St. Lawrence Seaway and Great Lakes*, by A. Cooperman, G. Cry, and H. Summer, 1959. A comprehensive report that includes such diverse subjects as cyclones, lake levels, and forecasting and warning services. Technical paper no. 35 of the U.S. Weather Bureau, Washington, DC.

Erosion and flooding

50. *Canada/Ontario Great Lakes Shore Damage Survey*, by Environment Canada and the Ontario Ministry of Natural Resources, edited by R. S. Boulden, 1975. This technical report summarizes the findings of a survey to ascertain the nature and extent of damages from shore erosion and flooding during the high lake level period November 1972 to November 1973. Published by Environment Canada, Ottawa, Ontario, and the Ontario Ministry of Natural Resources, Toronto, Ontario.

51. *Coastal Zone Atlas*, by Environment Canada and Ontario

Ministry of Natural Resources, edited by W. S. Haras and K. K. Tsui, 1975. A companion publication to reference 50. The most comprehensive atlas for the Canadian Great Lakes shore. This atlas portrays the erodible portion of the Canadian Great Lakes from Port Severn on Georgian Bay to Gananoque on the eastern end of Lake Ontario excluding the Niagara River. Shoreline stretches are shown on two sheets. Sheet 1 consists of three strips: photomosaic, township lot lines with 1973 shoreline, and histograms of recession or accession rates. Sheet 2 consists of five strips depicting shoreline damage, ownership, value, land use, physical characteristics, and existing protective works in damaged areas. Published by Environment Canada, Ottawa, Ontario, and the Ontario Ministry of Natural Resources, Toronto, Ontario.

52. *The 100-year Flood and Erosion Prone Area Maps and Guide*, by Environment Canada and Ontario Ministry of Natural Resources, 1978. These maps provide basic data for use in the planning and development of shoreland areas. They are primarily designed for use in planning in order to mitigate flood and erosion damage. Published by Environment Canada, Ottawa, Ontario, and the Ontario Ministry of Natural Resources, Toronto, Ontario.

53. *Great Lakes Region Inventory Report, National Shoreline Study*, by U.S. Army Corps of Engineers, 1971. A comprehensive report that includes shoreline maps showing erosion rates and flood-prone stretches. Report prepared by the North Central Division for the National Shoreline Study, House Document 93-121, U.S. 93rd Congress, 1st session, Washington, DC, pp. 157-97.

54. *U.S. Great Lakes Shore and Damage Study*, by the U.S. Army Corps of Engineers, 1978. This four-page summary gives shoreland damage findings (cost of protective measures and damages) for 1972-1976 by state and by lake, as well as a comparison to the 1951-1952 high lake level period. Published by the North Central Division, U.S. Army Corps of Engineers, Chicago, IL.

55. *Shore Erosion in Western Lake Erie*, by J. P. Coakley and H. K. Cho, 1972. Observations of lake processes, and the determination and interpretation of erosion rates were made from the Detroit River to Wheatley. Published by the International Association Great Lakes Research, proceedings of the 15th conference, Toronto, Ontario, pp. 344-60.

56. *The Effect of Storm Surge on Beach Erosion, Point Pelee* by J. P. Coakley, W. Haras, and N. Freeman, 1973. The November 14, 1972 storm surge at Point Pelee is discussed in terms of lake levels and erosion of the beach face. Published by the International Association Great Lakes Research, proceedings of the 16th conference, Toronto, Ontario, pp. 377-89.

57. *Application of ERTS-1 Digital Data to Water Transport Phenomena in the Point Pelee-Rondeau Area*, by R. P. Bukata, W. S. Haras, and J. E. Bruton, 1975. This paper uses satellite data and remote sensing methodologies to illustrate and evaluate the physical processes that have produced the Point Pelee and Pointe aux Pins landforms. Published by *Verhandlungen—*

Internationale Vereinigung fur theoretische und angewandte Limnologie, vol. 19, pp. 168–78.

58. *The Influence of Geology on Erosion Rates Along the North Shore of Lake Erie* by P. J. Gelinas and R. M. Quigley, 1973. An excellent study that relates erosion rates to wave energy and the nature of the bluff materials. Published by the International Association of Great Lakes Research, proceedings of the 16th conference, Toronto, Ontario, pp. 421–30.

59. *Proceedings: Workshop on Great Lakes Coastal Erosion*, edited by N. K. Rukavina, 1983, 1978, and 1976. These collected papers from the three workshops are technical and of greatest interest to scientists and students of the Great Lakes. The wide range of papers discuss coastal erosion, sediment transport, sedimentation, shore protection structures, and generally review studies going on within the Great Lakes. The 1983 edition is available from the Canadian National Research Council, Ottawa, Ontario K1A 0R6.

60. *Lakeshore Erosion—Lake Ontario from Niagara to Cobourg*, by G. B. Langford, 1949. This report covers field work done along the shore with particular emphasis on erosion problems, erosion rates, lake levels, lake currents, and erosion preventive measures. Published by the Ontario Ministry of Planning and Development.

61. *Bluff Slumping and Stability: A Consumer's Guide*, 1982, is one of a series of booklets, brochures, and guides published by the Michigan Sea Grant College Program. These helpful publications are designed specifically with the property owner in mind. This guide explains the causes of bluff instability and outlines control measures that might be taken including land-use controls, reshaping the bluff face, protection against wave attack, subsurface water drainage, and increasing slope stability through plantings. The 65-page publication is available from Michigan Sea Grant Publications Office, 2200 Bonisteel Boulevard, Ann Arbor, MI 48109.

In addition, several other Michigan Sea Grant publications addressing shore erosion and coastal hazards are available: *Shoreline Erosion: Questions and Answers*; *Deciding to Buy: Lakeshore Erosion Proceedings*; *The Michigan Shore Protection Demonstration Project: Final Report* (this is a technical report and of more specialized interest); *Shore Erosion: What to Do*; *If You Are Thinking about Buying Great Lakes Shoreline Property*; *Laboratory Investigation of Shore Erosion Processes*; *Low Cost Shore Protection on the Great Lakes: A Demonstration/Research Project*; *Coastal Engineering and Erosion Protection Report for the Year 1976–1977*; *An Investment Decision Model for Shoreland Protection and Management*; *Shoreline Erosion Special Problems for Realtors 1976–77*; *How Citizens Can Influence Legislation*.

62. *Lake Erie Shore Erosion, Lake County, Ohio: Setting, Processes, and Recession Rates from 1876 to 1973*, by C. H. Carter, 1976. One of a series of reports planned for each county along the Ohio shore of Lake Erie. Each report contains basic data on lake processes and the geologic setting as well as data on beaches, shore protection structures, and erosion rates. These

reports are a good place to start if you are considering buying property along the Ohio shore. Published by the Ohio Geological Survey as Report Investigation 99, Columbus, OH 43224.

63. *Lake Erie Shore Erosion and Flooding, Lucas County, Ohio*, by D. J. Benson, 1978. Report is similar to the Lake County report by Carter. This is Ohio Geological Survey Report Investigation 107, Columbus, OH 43224.

64. *Lake Erie Shore Erosion and Flooding, Erie and Sandusky Counties, Ohio: Setting, Processes, and Recession Rates from 1877 to 1973*, by C. H. Carter and D. E. Guy, Jr., 1980. Report is similar to Lake County report by Carter. This is Ohio Geological Survey Report Investigation 115, Columbus, OH 43224.

65. *Lake Erie Shore Erosion, Ashtabula County, Ohio: Setting, Processes, and Recession Rates from 1876 to 1973*, by C. H. Carter and D. E. Guy, Jr., 1983. Report is similar to Lake County report by Carter. This is Ohio Geological Report Investigation 122, Columbus, OH 43224.

66. *Guide to Lake Erie Bluff Stabilization*, 1984, is one of a series of booklets, brochures, and guides published by the Ohio State University Sea Grant Program. These helpful publications are designed specifically with the property owner in mind. This guide explains the causes of bluff erosion, and after explaining the importance of protecting the toe of the bluff from erosion, discusses ways of stabilizing the remainder of the bluff by reshaping the slope, dewatering, or by using vegetation. The 20-page publication is available from Ohio Sea Grant, 484 West 12th Avenue, Columbus, OH 43210.

In addition, several other Ohio Sea Grant publications dealing with shore erosion and coastal hazards are available: *Tips for Shoreline-Protection Structure Construction and Maintenance*; *Permits Required for Lake Erosion Abatement Works*; *Identifying Your Shoreline Erosion Problems*; *Lake Erie Shore Erosion*; *Beaches Are Shore Protection*; *Ashtabula County Coastal Tour—Drive It Yourself*; *The Lake County Coastline*; *Municipal Assistance in Financing a Group Erosion Abatement Project*.

67. *Shoreline Erosion and Flooding, Erie County, Erie, Pennsylvania*, by P. D. Knuth and G. R. Crowe, 1985. A comprehensive study of the Lake Erie shore in Pennsylvania that includes the nature of the shore materials, slope processes, and erosion rates. Published by the Great Lakes Institute, Erie, PA.

68. *Stratigraphy and Bluff Recession Along the Lake Erie Coast, New York*, by R. J. Geier and P. E. Calkin. A comprehensive survey of the nature of the shore deposits and recession rates along the New York shore (Chautauqua and Erie Counties), Lake Erie. Published by the New York Sea Grant Institute, Albany, NY 12246.

69. *Guide to Coastal Erosion*, by C. O'Neill, 1985 is one of a series of booklets, brochures, and guides published by the New York Sea Grant Extension Program. These helpful publications are designed specifically with the property owner in mind. This guide provides information on the major causes and types of coastal erosion. Included is a glossary of coastal erosion terms

and a checklist for diagnosing the cause of erosion at a specific site. The 20-page publication is available from New York Sea Grant Extension, Cornell University, Ithaca, NY 14853.

Several other New York Sea Grant publications dealing with shore erosion and coastal hazards are available: *Guidelines for the Effective Use of Floating Tire Breakwaters*; *Maintaining Coastal Erosion Control Structures*; *Developing New York's Great Lakes Coastline Roadmap* (a guide for putting together federal and state shore structure permit applications); *Questions to Ask before You Buy*; *Casualty Loss Tax Information for Coastal Property Owners*; *Controlling Bluff Groundwater along the Great Lakes*; *Guidelines for Selecting a Marine Contractor*.

Coastal engineering

70. *How to Protect Your Shore Property*, by the Ontario Ministry of Natural Resources (OMNR), 1986. The 63-page publication includes such topics as selecting a method of shore protection and alternative approaches. Emphasis is placed on the advantages and disadvantages of the different structures. Available from the OMNR Public Information Centre, Toronto, Ontario M7A 1W3.

71. *Help Yourself*, by the U.S. Army Corps of Engineers, no date. Brochure addresses erosion problems in the Great Lakes region. Of interest to coastal residents because it outlines coastal processes and illustrates a variety of shoreline engineering devices used to combat erosion. Free from the U.S. Army Corps of Engineers, North Central Division, 219 South Dearborn Street, Chicago, IL 60604.

72. *Effects of Large Structures on the Ohio Shore of Lake Erie*, by R. P. Hartley, 1964. A classic work on the effect of large man-made structures (mostly jetties) on erosion rates. Published by the Ohio Geological Survey as Report Investigations 53, Columbus, OH 43224.

73. *Shore Protection Structures: Effects on Recession Rates and Beaches from the 1870s to the 1970s along the Ohio Shore of Lake Erie*, by C. H. Carter, D. J. Benson and D. E. Guy, Jr., 1981. This study indicates that erosion rates have been reduced by shore protection structures. However, the beaches have become narrower and less continuous. Published in *Environmental Geology*, vol. 3, pp. 353–62.

74. *Dredging and Water Quality Problems in the Great Lakes*, by U.S. Army Corps of Engineers, 1969. The report evaluates the effects of dredging on water quality, including the effect of the disposal of dredged material in the open waters of the lakes. Published by the U.S. Army Corps of Engineers, Buffalo, NY 14207.

75. *Biotechnical Slope Protection and Erosion Control*, by D. H. Gray and A. T. Leiser, 1982. Although a technical reference, this book (271 pages) may be useful to property owners, especially those who own property bordered by bluffs, scarps, or slumping dune faces. This reference gives advice on the combined use of structures and plantings to stabilize slopes. The reference should be available through your library. Published by Van

Nostrand Reinhold Co., 135 West 50th Street, NY 10020.

76. *Beach Nourishment along the Southeast Atlantic and Gulf Coasts*, by Todd Walton and James Purpura, 1977. This article examines the success and failure of several beach nourishment projects. Published in *Shore and Beach*, July, pp. 10–18.

77. *Beach Behavior in the Vicinity of Groins*, by C. H. Everts, 1979. A description of the effects of two groin fields in New Jersey, which concludes that groins cause erosion in the downdrift shadow area. Published in the *Proceedings of the Specialty Conference on Coastal Structures*, no. 79, pp. 853–67, and available as reprint 79-3 from the U.S. Army Corps of Engineers, Coastal Engineering Research Center, Vicksburg, MS 39180.

78. *Low-Cost Shore Protection*, by the U.S. Army Corps of Engineers, 1982. A set of four reports written for the layperson under this title includes an introductory report, a property owner's guide, a guide for local government officials, and a guide for engineers and contractors. The reports are a summary of the Shoreline Erosion Control Demonstration Program and suggest a wide range of engineering devices and techniques to stabilize shorelines, including beach nourishment and vegetation. In adopting these approaches, you should keep in mind that they are short-term measures and may have unwanted side effects. The reports are available from Section 54 Program, U.S. Army Corps of Engineers, USACE (DAEN-CWP-F), Washington, DC 20314.

79. *Shore Protection Manual*, by the U.S. Army Corps of Engineers, 1984. The bible of shoreline engineering. Published in two volumes. Request publication 008-022-00218-9 from the Superintendent of Documents, U.S. Government Printing Office, Washington, DC 20402.

80. *Publications List, Coastal Engineering Research Center (CERC) and Beach Erosion Board (BEB)* by the U.S. Army Corps of Engineers. A list of published research by the U.S. Army Corps of Engineers. Free from the U.S. Army Corps of Engineers, Coastal Engineering Research Center, Vicksburg, MS 39180.

Site analysis and home construction

Current and prospective owners and builders of homes in flood-prone areas should supplement the information and advice provided in this book with that offered in references dealing specifically with safe construction. These excellent references contain sound, useful information that should help the residents of such areas minimize the losses caused by high wind or rising water. Many of these publications are free. The government publications are paid for by your taxes, so why not use them? The following references are recommended to those readers who wish to further investigate the subject of flood-resistant construction.

81. *Standard Building Code* is available from the Southern Building Code Congress, 1116 Brown Marx Building, Birmingham, AL 35203.

82. *The BOCA Basic Building Code* is available from the Building Officials and Code Administrators International, Inc.,

17926 South Halsted St., Homewood, IL 60430.

83. *The Uniform Building Code* is available from the International Conference of Building Officials, 5360 South Workman Mill Road, Whittier, CA 90601.

84. *Manufactured Home Installation in Flood Hazard Areas*, by the Federal Emergency Management Agency, 1985. This book outlines methods for tying down mobile homes, elevating techniques, foundations, bracings, and related methods of floodproofing such structures. Useful appendixes are included. Available from FEMA offices (see appendix A) or the U.S. Government Printing Office, Washington, DC 20402. Request publication no. 1985-529-684/31054.

85. *Interim Guidelines for Building Occupant Protection from Tornadoes and Extreme Winds*, TR-83A, and *Tornado Protection — Selecting and Designing Safe Areas in Buildings*, TR-83B. These publications are available from the Civil Preparedness Agency, Department of Defense, the Pentagon, Washington, DC 20301; or from the Civil Defense Preparedness Agency, 2800 Eastern Boulevard, Baltimore, MD 21220.

86. *Economic Feasibility of Floodproofing — Analysis of a Small Commercial Building*, by the Federal Emergency Management Agency, 1979. A riverine case study from Pennsylvania provides a model that is applicable to the Lake Erie shore and floodplains of rivers draining into the lake. Available through FEMA offices (see appendix A) or from the U.S. Government Printing Office, Washington, DC 20402. Request publication no. 1979-0-628-156/2005 Region 3-1.

87. *A Coastal Homeowner's Guide to Floodproofing*, by the Massachusetts Disaster Recovery Team. This booklet combines a checklist to take the home owner through the process of floodproofing existing houses with tips on dealing with engineers and contractors. Suggestions and guidelines are appropriate to houses in flood zones along the Great Lakes. Available from the Office of the Lieutenant Governor, State House, Boston, MA 02133.

88. *Coastal Design: A Guide for Planners, Developers, and Home Owners*, by Orrin H. Pilkey, Jr., Orrin H. Pilkey, Sr., Walter D. Pilkey, and William J. Neal, 1983. A detailed companion volume to the "Living with the Shore" series and a construction guide expanding on the information outlined in this text. Chapters include discussions of shoreline types, individual residence construction, making older structures storm worthy, high-rise buildings, mobile homes, coastal regulations, and the future of the coastal zone. Published by Van Nostrand Reinhold, 135 West 50th Street, New York, NY 10020.

89. *Design and Construction Manual for Residential Buildings in Coastal High Hazard Areas*, prepared by Dames and Moore for the Department of Housing and Urban Development on behalf of the Federal Emergency Management Agency (FEMA), Federal Insurance Administration, 1986. A guide to the coastal environment with recommendations on site and structure design relative to the National Flood Insurance Program. The report includes a model building code, design considerations, examples, construction costs, and appendixes on design tables, bracing, design worksheets, wood preservatives, and a list of

useful references. Available from the Superintendent of Documents, U.S. Government Printing Office, Washington, DC 20402, or any FEMA office (see appendix A).

90. *Elevated Residential Structures*, prepared by the American Institute of Architects Foundation, 1984. An excellent outline of the flood threat and the necessity for proper planning and construction. The manual is limited to the special design issues confronted in elevation construction. Illustrates construction techniques and includes glossary, references, and worksheets for estimating building costs. The readers of this manual are assumed to have knowledge of conventional residential construction practice. Order publication 0-222-193 from the Superintendent of Documents, U.S. Government Printing Office, Washington, DC 20402, or contact a FEMA office.

91. *Flood Emergency and Residential Repair Handbook*, prepared by the NAHB Foundation, Inc., of the National Association of Home Builders, 1986. Guide to floodproofing as well as step-by-step cleanup procedures and repairs including household goods and appliances. Available from your FEMA office (see appendix A). Order publication no. FIA-13.

Legislation

The following are specific references to the federal legislation mentioned in chapter 5. The first reference indicates where these can be found in the *United States Code*. The second refers to their place in *Statutes at Large*.

92. National Flood Insurance Act of 1968 (P.L. 90-448), enacted on August 1, 1968. (1) 42 U.S.C. sect. 4001 et seq. (1976); (2) 82 Stat. 476, Title 13.

93. Coastal Zone Management Act of 1972 (P.L. 92-583), enacted on October 27, 1972. (1) 16 U.S.C. sect. 1451 et seq. (1976); (2) 82 Stat. 1280. This act was amended on January 2, 1975 (P.L. 93-612. (1) 16 U.S.C. sects. 1454, 1455, and 1464; (2) 88 Stat. 1974); and on July 26, 1976 (Coastal Zone Management Act Amendments of 1976, P.L. 94-370. (1) 5 U.S.C. sect. 5316, 15 U.S.C. sect. 1511a, and 16 U.S.C. sects. 1451 and 1453-1464; (2) 90 Stat. 1013).

94. Federal Water Pollution Control Act Amendments of 1972 (P.L. 92-500), enacted on October 18, 1972. (1) 33 U.S.C. sect. 1251 et seq. (1976); (2) 86 Stat. 816.

95. Flood Disaster Protection Act of 1973 (P.L. 92-234), enacted on December 31, 1973. (1) U.S.C. sect. 4001 et seq. (1976); (2) 87 Stat. 975.

96. Water Resources Development Act of 1974 (P.L. 92-251), enacted on March 7, 1974. (1) 16 U.S.C. sects. 4601-13, 4601-14, and 460ee (1976); 22 U.S.C. sect. 275a (1976); 33 U.S.C. sects. 59c-2, 59k, 579, 701b-11, 701g, 701n, 701r, 701r-1, 701s, 709a, 1252a, and 1293a (1976); 42 U.S.C. sects. 1962d-5c, 1962d-15, 1962d-16, and 1962d-17 (1976); (2) 88 Stat. 13, Title 1.

97. Coastal Barrier Resources Act (P.L. 97-348), enacted on October 18, 1982. (1) 16 U.S.C. sect. 3501 et seq. (2) 96 Stat. 1653.

98. *How to Read Flood Hazard Boundary Maps*, by the

National Flood Insurance Program, 1981. This foldout pamphlet provides a step-by-step guide to reading the FHBM. A similar pamphlet, *How to Read a Flood Insurance Rate Map*, 1981, provides a guide for citizens, officials, lending institutions, and insurance agents. Request publications FIA-3 and FIA-10 from your FEMA office (see appendix A).

99. *Ocean and Coastal Law*, by Richard Hildreth and Ralph Johnson, 1983. Discusses (1) problems posed by earlier nonmanagement of public resources and the gradual public awakening to the need for a comprehensive legal framework for the coastal zone, (2) ownership and boundary questions, (3) state common law, (4) offshore issues, (5) alteration of waterways and wetlands, and (6) federal and state coastal zone management programs. For anyone with a serious interest in legal issues. Published by Prentice-Hall, Englewood Cliffs, NJ 07632.

Planning and management

100. *Regulation of Great Lakes Water Levels*, by the International Great Lakes Levels Board, 1973. One of the most comprehensive reports on the question of lake level regulation. The report includes individual appendixes on: hydrology and hydraulics (A), lake regulation (B), shore property (C), fish, wildlife, and recreation (D), commercial navigation (E), power (F), and regulatory works (G). For information about the report, contact the International Joint Commission, 100 Ouellette Avenue, Windsor, Ontario N9A 6T3.

101. *Lake Erie Water Level Study* by the International Lake Erie Regulation Study Board, 1981. This report, while similar to the 1973 study, focuses on limited regulation of Lake Erie. Because of the interrelated nature of the lakes, the effects of Lake Erie regulation are determined for all the lakes. The report includes individual appendixes on: lake regulation (A), regulatory works (B), coastal zone (C), commercial navigation (D), power (E), environmental effects (F), recreational beaches and boating (G), and public information programs (H). This study is one of the best summaries on lake level regulation. For information contact the International Joint Commission, 100 Ouellette Avenue, Windsor, Ontario N9A 6T3.

102. *National Shoreline Study—Shore Management Guidelines*, by U.S. Army Corps of Engineers, 1971. The report includes information and suggestions for preserving, conserving, and enhancing the shoreline. Published by the U.S. Army Corps of Engineers, Section 54 Program, Washington, DC 20315.

103. *Great Lakes Basin Framework Study*, 1975. The now disbanded Great Lakes Basin Commission published a *Report of the Great Lakes Basin Framework Study* and 25 appendixes to accompany the report. The appendixes are valuable sources of information for Lake Erie as well as the other lakes. Appendixes most pertinent to the Lake Erie shore include: geology and groundwater (3), limnology of lakes and embayments (4), levels and flows (11), shore use and erosion (12), land use and management (13), and erosion and sedimentation (18). Most libraries in the Great Lakes region have a set of the report and its appendixes.

104. *Assessment of flood and Erosion Assistance Programs: Rondeau Coastal Zone Experience, Lake Erie*, by J. C. Day, J. A. Fraser, and R. D. Kreutzwiser, 1977. An evaluation of federal and provincial assistance programs in the Rondeau area in the early 1970s. Suggestions are given for future programs. Published in the *Journal of Great Lakes Research*, vol. 3, no. 1–2, pp. 38–45.

105. *Coastal Processes and Shoreline Encroachment: Implications for Shoreline Management in Ontario*, by R. Davidson-Arnott and R. Kreutzwiser, 1985. The authors stress the need for a commitment to coastal zone management for a long-term approach to the problems of flooding and shore erosion. Published in the *Canadian Geographer*, vol. 29, no. 3, pp. 256–62.

106. *An Evaluation of Government Response to the Lake Erie Shoreline Flood and Erosion Hazard*, by R. Kreutzwizer, 1982. The author states that there is too great a reliance, both in Canada and the United States on federal support of shore protection structures and continued development of shoreland. Published in the *Canadian Geographer*, vol. 26, no. 3, pp. 263–73.

107. *Economic Portrait of Great Lakes Region*, by the Federal Reserve Bank of Chicago in cooperation with the Great Lakes Commission, 1986, is a statistical compendium compiled for individual Great Lakes states. Available from Harbor House Publishers, Boyne City, MI.

108. *The Faces of the Great Lakes*, by B. A. King, photographer, and Jonathan Ela, writer, 1977, is a coffee-table photo album with a message. The photo essay captures the variety of the Great Lakes. The text directly addresses the important issues of lakeshore planning, conserving rural shorelines, public access, water quality, the fisheries, and saving the lives of the Great Lakes. Published by Sierra Club Books, San Francisco, CA 94108, and available through your bookstore or library.

109. *Coastal Design with Natural Processes along the Ocean and Great Lakes Coastlines of the U.S.—An Introduction*, by S. H. Lopez, 1985. A paper that discusses how best to use our coastal areas in an environmentally sensitive and rational fashion. Published in *Landscape Architecture Technical Information*, series 9, November 1985.

110. *Development of Coastal Resources Proximate to the Port of Buffalo*, by R. F. Paaswell, W. W. Recker, and A. F. Brundage, 1979. Published by the New York Sea Grant Institute, Albany, NY.

111. *Beach Erosion Control District Feasibility Study*, by Great Lakes Laboratory, State University College at Buffalo, 1980. The study, which includes seven volumes with titles ranging from *Literature Review* to *Financing and Projected Costs*, is for Erie County, New York. The work provides a working model for other shoreline areas along the Great Lakes. Available from the New York Department of State, Albany, NY.

112. *Guidelines for Local Waterfront Revitalization Programs* by New York Department of State, 1982. The report covers topics that range from an inventory analysis to public access policies. Available from Department of State, Albany, NY.

113. *Second Skidaway Institute of Oceanography Conference*

on *America's Eroding Shoreline*, a white paper by J. D. Howard and others, 1985. This conference produced recommendations that a plan be developed at different levels of government to move development back from retreating shorelines. This conference was a follow-up to a first conference: "Saving the American Beach: A Position Paper by Concerned Coastal Geologists," 1981. Subtitled "National Strategy for Beach Preservation," most of the document's recommendations are appropriate for the Great Lakes. Both white papers are available from the Director, Skidaway Institute of Oceanography, P. O. Box 13687, Savannah, GA 31416.

114. *The Beaches Are Moving: The Drowning of America's Shoreline*, by Wallace Kaufman and Orrin Pilkey, 1979. This highly readable account of the state of America's coastline explains natural processes at work at the beach, provides a historical perspective of man's relation to the shore, and offers practical advice on how to live in harmony with the coastal environment. Originally published by Anchor Books-Doubleday, it is now available in paperback with a 1983 epilogue by the authors from Duke University Press, 6697 College Station, Durham, NC 27708.

115. *Living with the New Jersey Shore*, by Karl F. Nordstrom and others, 1986. A volume in the Living with the Shore series, which provides examples of the end point of shoreline stabilization: massive seawalls and groins without beaches. Some of America's once-famous resort beaches have been lost as a result of holding a migrating shoreline in place. Available through

bookstores, or Duke University Press, Durham, NC 27708.

116. *Natural Hazard Management in Coastal Areas*, by Gilbert White and others, 1976. Summary of coastal hazards along the entire coast of the United States. Discusses adjustments to hazards and hazard-related federal policy and programs. Summarizes hazard management and coastal planning programs in each state. Appendixes include a directory of agencies, an annotated bibliography, and information on storms. A valuable reference, recommended to developers, planners, and managers. Available from the Office of Coastal Zone Management, National Oceanic and Atmospheric Administration, 3300 Whitehaven Street, N.W., Washington, DC 20235.

117. *Coastal Flood Hazards and the National Flood Insurance Program*, prepared by H. Crane Miller for the Federal Insurance Administration, 1977. Reports the findings of a study undertaken to determine the effects of the National Flood Insurance Program on 15 selected Atlantic and Gulf coastal areas. Available from the U.S. Government Printing Office, Washington, DC 20402. Ask for publication 721-300/734.

118. *Questions and Answers on the National Flood Insurance Program*, by the Federal Emergency Management Agency (FEMA), 1983. Pamphlet explaining the basics of flood insurance and providing addresses of FEMA offices. Free from the Federal Emergency Management Agency, Washington, DC 20472.

119. *Development of the Coast: Facing the Tough Issues*, a Coastal States Organization (CSO) conference held in Charleston, 1979. The published proceedings of this conference give an

abbreviated overview of the wide range of problems generated by coastal development. Available from the CSO, Conference Management Associates, 1044 National Press Building, Washington, DC 20045.

120. *Who's Minding the Shore?* by the Natural Resources Defense Council, 1976. A guide to public participation in the coastal zone management process. Defines coastal ecosystems and outlines the Coastal Zone Management Act, coastal development issues, and means of citizen participation in the coastal zone management process. Lists sources of additional information. Available from the Office of Coastal Zone Management, National Oceanic and Atmospheric Administration, 3300 Whitehaven Street, N.W., Washington, DC 20235.

121. *The Fiscal Impact of Residential and Commercial Development, A Case Study*, by T. Muller and G. Dawson, 1972. A classic study that demonstrates that development may ultimately increase, rather than decrease, community taxes. Available from the Publications Office, the Urban Institute, 2100 M Street, N.W., Washington, DC 20037. Refer to URI-22000 when ordering.

122. *Coastal Ecosystem Management*, by John Clark, 1977. This 928-page text covers most aspects of the coastal zone from descriptions of processes and environments to legal controls and outlines for management programs. Essential reading for planners and beach community managers. Published by John Wiley and Sons, 605 Third Avenue, New York, NY 10158. Available in most university libraries.

123. *Coastal Environmental Management*, prepared by the Conservation Foundation, 1980. Guidelines for conservation of resources and protection against storm hazards, including ecological description and management suggestions for coastal uplands, floodplains, wetlands, banks and bluffs, dunelands, and beaches. Part 2 presents a complete list of federal agencies and their authority under the law to regulate coastal zone activities. A good reference for planners and persons interested in good land management. Available from the Superintendent of Documents, U.S. Government Printing Office, Washington, DC 20402.

Newsletters and directories

124. *The Great Lakes Reporter*, by the Center for the Great Lakes, is a bimonthly newsletter for the Great Lakes. It reports on significant programs and policies taking place on the Lakes. Available from the Center for the Great Lakes, 435 N. Michigan Avenue, Suite 1733, Chicago, IL 60611.

125. *Focus on Great Lakes Water Quality*, by the International Joint Commission (IJC), is a quarterly report on topics related to water quality. Includes sections that range from upcoming events to book reviews. Available from the IJC, 100 Ouellette Avenue, Windsor, ONT N9A 6T3.

126. *Upwellings*, by the Michigan Sea Grant College Program, is a quarterly newsletter focusing on Great Lakes topics including issues, summaries of research, announcements of

events and publications, and similar educational and advisory news. Subscription and program information is available from the Michigan Sea Grant College Program, Ann Arbor, MI 48109.

127. *Water Impacts*, by the Institute of Water Research, is a monthly newsletter on water issues and information including the Great Lakes. Contact *Water Impacts*, Institute of Water Research, Michigan State University, East Lansing, MI 48824.

128. *Middle Sea*, by the Ohio Sea Grant, is a newsletter for Ohio educators. The periodical includes school activities, instructional articles, issues in the news, and announcements of relevant events that center around Lake Erie or the Great Lakes. Copies and information are available from the Ohio Sea Grant Education Program, The Ohio State University, Columbus, OH 43210.

129. *Coastlines*, by New York Sea Grant Extension Program, is a bimonthly newsletter aimed at people who work and play in the coastal zone. Available from New York Sea Grant Extension, Cornell University, Ithaca, NY 14853.

130. *The Great Lakes Directory of Natural Resource Agencies and Organizations*, edited by Paula Ripley and produced by the Freshwater Society, 1984. This directory provides an exhaustive listing of U.S. and Canadian federal, state, provincial, local, public, and private agencies with interest and activities related to the Great Lakes and their drainage basin. In addition to consolidating addresses and phone numbers, this useful resource volume states each agency's purpose, activities, staff size, programs,

publications, and related information. Its 212 pages are packed with information as to who is doing what in the Great Lakes region. Available from the Center for the Great Lakes, 435 North Michigan Ave., Suite 1733, Chicago, IL 60611.

New FEMA publication

As this book went to press, FEMA issued a new floodproofing guide:

131. *Floodproofing Non-Residential Structures*, by Booker Associates, Inc. for the Federal Emergency Management Agency, 1986. This book outlines factors that influence floodproofing measures (including specific site factors), design of floodproofing measures, cost and benefits of floodproofing, and examples of typical applications. The five appendixes include a glossary, lists of useful publications and agencies, floodproofing performances criteria, and a guide to determining flood depths using flood insurance maps. Available from FEMA offices (see appendix A). Request publication no. FEMA 102/May 1986.

Index

About the Authors

Charles H. Carter is Associate Professor of Geology, the University of Akron, and is a coastal geologist with many years experience in studying the Lake Erie shore, including ten years with the Ohio Geological Survey. William Haras is an engineer, hydrographer, and planner now with the Canada Centre for Inland Waters, Burlington, Ontario; he has worked on several water-level studies for the International Joint Commission for the Great Lakes. William J. Neal is Professor of Geology, Grand Valley State College, Allendale, Michigan. Orrin H. Pilkey, Jr., is James B. Duke Professor of Geology at Duke University.